AIRSTREAM COUNTRY

AIRSTREAM COUNTRY

A Geologic Journey Across the American West

NEIL MATHISON

UNIVERSITY OF NEW MEXICO PRESS | ALBUQUERQUE

Library of Congress Cataloging-in-Publication Data
Names: Mathison, Neil, author. Title: Airstream country: a
geological journey across the American West / Neil G. Mathison.
Description: Albuquerque: University of New Mexico Press, 2024.
Identifiers: LCCN 2024023912 (print) | LCCN 2024023913 (ebook)
| ISBN 9780826367464 (paperback) | ISBN 9780826367471 (epub).
Subjects: LCSH: Mathison, Neil—Travel—West (U.S.) | Retirees—
Biography—21st century. | Geology—West (U.S.) | West (U.S.)—
Description and travel. | LCGFT: Travel writing. Classification:
LCC PS3613.A8288 A3 2024 (print) | LCC PS3613.A8288 (ebook) |
DDC 814.6—dc23/eng/20240612. LC record available at https://lccn.
loc.gov/2024023912. LC ebook record available at https://lccn.loc.
gov/2024023913

Founded in 1889, the University of New Mexico sits on the traditional homelands of the Pueblo of Sandia. The original peoples of
New Mexico—Pueblo, Navajo, and Apache—since time immemo-
rial have deep connections to the land and have made significant
contributions to the broader community statewide. We honor the
land itself and those who remain stewards of this land throughout
the generations and also acknowledge our committed relationship
to Indigenous peoples. We gratefully recognize our history.

Cover illustration courtesy of Tyler Casey on Unsplash
Designed by Felicia Cedillos
Composed in Alegreya

To Susan. Without Susan, no book.

CONTENTS

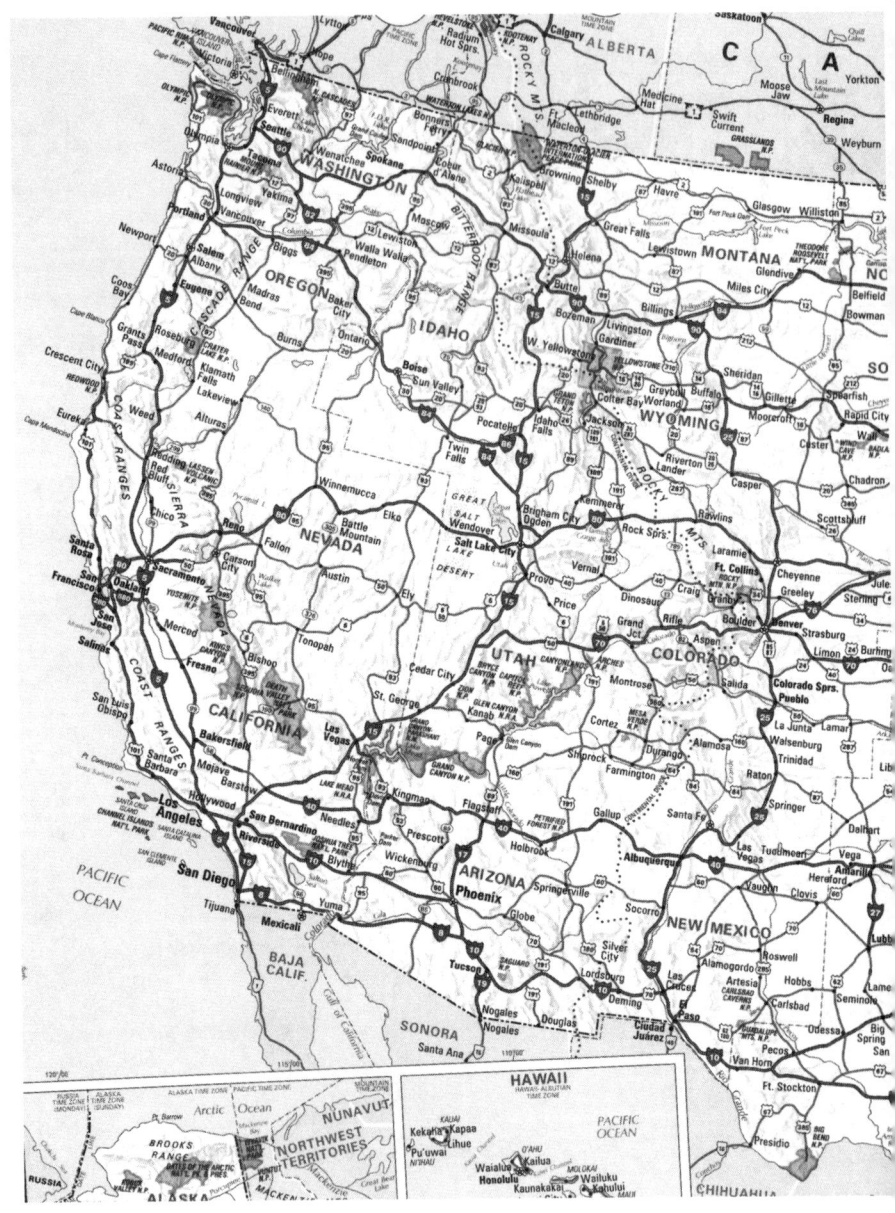

GENESIS

IN SEPTEMBER 2011 MY wife, Susan, retired from Microsoft. She was fifty-five years old, in good health, mother to John, our son, who was then attending college in the San Francisco Bay Area. She had worked full time for over thirty years, beginning during her undergraduate study at the University of Washington, while earning a BS in microbiology and a BA in zoology. Later, still full-time employed, she would earn an MBA. "I need an event," Susan announced, "to mark the change." The event she chose was to raft down the Grand Canyon in human-propelled rubber rafts and wooden dories.

Within seventy-two hours of her last day at Microsoft, Susan, I, and nineteen other soon-to-be rafters sat in an outdoor bar drinking gin and tonics below Utah's Vermilion Cliffs, above the Colorado River's Marble Canyon, where, the next morning, at Lees Ferry, we would embark on a river journey through the Grand Canyon. We would run rapids, hike side canyons, endure monsoonal rains ("a once-in-a-century event," according to our guides), camp in tents or under the stars on riverbank sandbars and under hollows in the canyon wall. The course of our days would be determined by "River Time," marked by the rising and setting of the sun, by what campsites were available, by the miles the river carried us each day. We had no commitments, no obligations, no cell phones. On our egress from the canyon nineteen days later, the rafting company would bus us back to Las Vegas. Susan's retired life—*our* retired life—would begin.

Because we had no obligations (we were retired!), we tent-camped home to Seattle via the Sequoia and Kings Canyon National Parks, the redwoods, and the Oregon coast. When the rain began to fall in Newport, Oregon, we decided to forgo our tent and check into a motel. When it rained at Cannon Beach, we did the same. Could it be that we needed an alternative to a tent?

En route to Seattle, we stopped at Portland, Oregon's Airstream Northwest travel-trailer dealer. A used sixteen-foot Airstream Bambi trailer was for sale. Just the one we wanted. We sat in our car in the dealer's parking lot and later in a restaurant, trying to decide whether to buy it. The price, by Airstream standards, was fair. And we knew that used Airstreams were rarely on the market. By lunch's end, however, we decided that too many things needed to be unwound before we threaded one into our lives: John's education, a Boston Whaler boat to be sold, a winter lived and skied in Sun Valley.

I had made a major life change, too, twelve years earlier. After graduating from the US Naval Academy, I had, like Susan, worked one job after another: ten years as a nuclear propulsion–trained naval officer, five years as a computer salesperson and sales manager in Seattle, six years in Hong Kong as general manager for two different tech company subsidiaries, and five more as a sales vice-president at two corporate headquarters, one in Palo Alto, the other in the Seattle area. I needed a break. Or maybe I needed an entirely new career. I also wanted an event to mark the milestone, so I rented a recreational vehicle. Then, with son John, at the time five years old, and John's nanny, Vilma, and with Susan flying in on weekends, we made a three-week spring break loop through the Rocky Mountains and American Southwest.

In the next dozen years, between summer sailing vacations and winter skiing, we managed a half-dozen car-and-tent camping trips, usually during John's spring breaks, sometimes in June after he got out of school. Our goal was to take our son to visit the major national parks and monuments in the western United States and Canada: Yellowstone, Yosemite, and Grand Canyon; Glacier, Carlsbad Caverns, Rocky Mountain, and Grand Teton; Bryce, Zion, and Monument Valley; Banff, Lake Louise, and Waterton Lakes. Unlike rafting the Grand Canyon, we didn't live by "River Time." We were in the middle years of busy lives: Susan's work, John's school, my efforts to become a writer. Professional, academic, and recreational obligations governed us.

By the spring of 2013, however, our retired life had taken shape. Winter in Idaho's Sun Valley. Summers on Washington State's San Juan Island. Augusts on our sailboat in British Columbia. John's college over. It was time to find that alternative to a tent.

In May 2013, in Covington, Washington, south of Seattle, we took delivery of a brand-new sixteen-foot Airstream Bambi travel trailer. Our Bambi Diaries had begun.

CHAPTER 1

SINEWS OF A CONTINENT

It is better to travel well than to arrive.

—BUDDHA

IT'S MORNING, SIX O'CLOCK, in Northeastern Oregon, at Farewell Bend State Park. In a few minutes, I'll roll out of the Airstream's double bed, light the propane stove, draw water for my coffee and my wife Susan's tea, pull on a sweater to ward off the morning chill, and greet the sun rising over the Snake River. If you don't mind small spaces—our Bambi is sixteen feet long, hitch to taillights—it's lovely in here. An aluminum cocoon with no sharp corners and few rectangular planes. The light here suffuses and softens, like being inside a snow globe or a Zen meditation.

At this point, Susan and I have owned our Airstream Bambi for fourteen months. We're near the end of our second long trip: a loop from western Washington across the North Cascades, into Canada's Okanogan Valley, then along the bottom of British Colombia, up into the Canadian Rockies, over the Continental Divide, and south to Canada's Waterton Lakes National Park, reentering the United States at Glacier National Park. We passed through a covey of creepy fenced-off Air Force nuclear-missile silos, whether active or mothballed we don't know. Then we skipped Montana's capital, Helena, and its most famous mining town, Butte, and headed southwest into Idaho. After a few days' respite in Ketchum, site of the Sun Valley Ski Resort, we've set out again on the home stretch, at least for this trip.

I swing my feet from under the quilt, rise up, face the stove. From where I stand, I can touch all the primary appurtenances: stove, fridge, sink, toilet door, dinette, overhead cabinets, air conditioner, heater, even sleeping Susan's shoulder. I slide the front curtains open. The Snake runs less than fifty feet below us, a glittering gold braid in the rising sun. Because it's dammed downstream, it flows slowly here, even though it's late June and still fed by spring runoff—snowmelt from the Tetons and the Northern Rockies. In the nineteenth century, Oregon Trail pioneers camped here, bound for the Willamette Valley. At Farewell Bend, they bid goodbye to the water-and-life-giving Snake and began the dry, high-desert trek across the top of central Oregon.

I unlatch the Bambi's airplane-style door, which curves top to bottom to fit the Airstream's aerodynamic shape. The fragrance of sagebrush, cut grass, and ponderosa pine wafts in. Semitrailer trucks drone along nearby Interstate 84, on the Snake's southern shore. On the opposite, Idaho shore, the Burlington Northern Santa Fe railway parallels the river—last night several trains woke us as they clattered by. Before I-84, there was US 30, a road I traveled in my pre–interstate highway boyhood. Before US 30, there was the Oregon Trail.

Since we began our Bambi travels, friends never ask what they want to ask: *So many miles? So small a space?*

What exactly, they wonder, do we intend by Bambi travel?

We would answer this: Nostalgia. Serendipity. Simplicity. Science.

Susan and I both grew up in the West. We want to go where we've been but also where we haven't. We want to travel with less haste and less structure than we did in our pre-retirement past and to live a minimalist life. We want to understand the bones of the West: how it connects, how its geology and its geography came to be.

There's much to be said for road travel. You see more than you see from airplanes. You have more choices than you do by train. You cover more miles than you can by walking or riding a bike. And if you want a sense for a large place like the West, it can be understood best when you scale its mountain ranges, trace its rivers, explore its canyons, follow its coasts—and the best way to do that, perhaps the only time-efficient way, is to drive its highways. Highways are the West's sinews. They connect us. When you travel by car, you see the transitions from one place to another. You fill in the lacuna of the land—deserts to mountains, mountains to prairie, prairie to coast.

Maybe that's what Susan and I want: to touch the sinews of the land.

Or maybe it's just this: The open road calls, as it has to so many others: Kerouac in *On the Road*, Steinbeck in *Travels with Charley*, Wolfe in *The Electric Kool-Aid Acid Test*, William Least Heat-Moon in *Blue Highways*, Robert Pirsig in *Zen and the Art of Motorcycle Maintenance*, Frodo Baggins in *The Fellowship of the Ring*.

Maybe that's what we're doing: answering the call of the road.

Coffee brewed, I step outside and seat myself in one of our folding camp chairs. Why does coffee taste so much better out of doors? The morning sun warms my face; in a few hours it'll be pushing a hundred degrees. The terrain here is desert-dry. Hills rise steeply on both sides of the river, covered by sagebrush and rabbit brush and an occasional juniper or piñon pine. Locals call this "high desert" but a more accurate name may be "shrub steppe," *steppe* being a Russian word forged in the open spaces of Russia, Uzbekistan, and Kazakhstan. In the summer, this place is hot with little rain, in the winter, cold with relatively light snowfall. The steppe. Kazakhstan, northeastern Oregon. Our place in the world.

I hear Susan stir. Soon she joins me, seating herself in our other folding chair. We discuss our travel goal for the day. We'll go all the way, we decide, to western Washington's Skagit Valley, where we intend to store the Bambi during our summer sailing trip. For the moment, however, we just savor here. The sun. The river flowing. Coffee and tea. Tomorrow there will be a different view, a different place.

Susan grew up a tent camper but she's new to trailering. My parents purchased their first travel trailer in 1950. I was three years old. The trailer was homemade (although not by their hands), fifteen feet long, and tall enough for an adult to stand up in, but lacked a stove or toilet. It had a full-breasted, windowless front that swept up and back and down to the trailer's rear like a ducktail haircut, framed in spruce and sided with plywood under thick, layered canvas. My parents painted it green on the bottom, belted it with a red stripe and topped it with gray, and named it Papagayo, after a parrot they'd owned in Brazil that had died from drinking too much Coca Cola. We spent most weekends each year, March to October, in Papagayo.

Summers, during my father's three-week vacation, we traveled the American, Canadian, and Mexican West's most notable places: Yellowstone, Glacier, the Grand Tetons, Yosemite, Mount Lassen, Crater Lake, Banff, Lake Louise, Waterton Lakes, the Grand Canyon, Bryce Canyon, Zion, Craters of the Moon, Arches, Mesa Verde, Lava Beds, Death Valley, Sequoia/Kings Canyon, Organ Pipe, Saguaro, Tucumcari, Carlsbad Caverns, Great Sand Dunes, the Rocky Mountains, San Francisco, Los Angeles, San Diego, Denver, Salt Lake City, Albuquerque, Tucson, Phoenix, Las Vegas, Guaymas, Mazatlán, Guadalajara. What we did was more than sightseeing. We were birthing family legends: my father backing Papagayo two miles up a winding coastal track perilously teetering above the Pacific Ocean because there was no room at the end-of-road campsite to turn around; a hitch ball snapping on a Friday-evening run to Mount Rainier (a new ball purchased at closing time from a Chubby and Tubby in south Tacoma); my brothers and me contracting measles at Banff and Lake Louise during a Canadian Rockies vacation; a wheel spinning off Papagayo along Idaho's US 30, the axle screeching across the highway asphalt, my father veering us to the shoulder, the wheel bouncing into the desert sagebrush. Harrowing mountain passes became part of our white-knuckle familial lore: Farmington Flats in the Wasatch Mountains behind the Great Salt Lake, Sonora Pass over the Sierra Nevada.

My parents left us no real estate, no stocks and bonds, no cash. But my brothers and sister and I agree: the American West was gift enough.

Our goal today is Bay View State Park in the Skagit Valley on Samish Bay in Washington State, near what will be the Bambi's July and August home. It's a long run, so we have to leave early. We've become adept at early departures. Before we leave, we fill the trailer water tanks. We unscrew and roll up the hose, unplug and coil the electrical cord, and store both in the trailer's tail-end compartment. We fold up the camp chairs, raise the stabilizing jacks (one in each of the four corners to keep the Bambi from pitching about when we're parked), hitch the car to the trailer, and remove the trailer wheel chocks. We secure the Bambi's interior so that plates, glasses, DVDs, pots, pans, socks, T-shirts, shoes, books, wine bottles, and underwear don't fly about when we're underway. We adjust the trailer-towing mirrors (which extend on metal arms from the regular car side mirrors), test the trailer taillights, and make certain all of the trailer vents and windows are closed.

A few miles west of Farewell Bend, we pass from the Mountain Time Zone into the Pacific Time Zone. The highway follows the Burnt River here, more a creek than a river. We pass through limestone hills with quarries, retired and active. Once, 150 million to 200 million years ago, this was a shallow sea. Now it's bone-dry desert. We follow I-84 through the Blue Mountains of Oregon, across two glorious mountain-ringed valleys, the first where Baker City is located, the hometown of Airstream founder Wally Bynum, the second where La Grande is sited. A road sign informs us that we've crossed the latitude of 45 degrees north and are halfway between the equator and the North Pole. We summit the Blue Mountains at about 4,500 feet, then descend Emigrant Hill to Pendleton, site of the Pendleton Round-Up rodeo, intersect with I-82 at the Umatilla Ordnance Depot, where the US Army destroyed much of its chemical weapons stockpile after the Cold War, cross the Columbia River into Washington State, and follow I-82 through the Yakima Valley to the city of Ellensburg. There we join Interstate 90 to Seattle and navigate the city traffic before heading north. At Mount Vernon, in the Skagit Valley, we head west to Bay View State Park.

The park lies forty miles south of the USA–Canada border in the Skagit River Valley delta. Its fronts the Padilla Bay Estuarine Natural Reserve, a wetlands and avian flyway on the delta's western littoral. Douglas fir–topped hillocks rise from a mostly flat valley. Without the Skagit River delta, the hills might have become islands in Puget Sound. Looking north, we see the snow-topped British Columbia Coastal Range. To the southwest is the zigzag silhouette of the Olympic Mountains. To the east rise the volcanic cone of Mount Baker and the snowy Cascade summits. The valley floor is layered with fertile glacial till ground from Cascade Mountain basalt, volcanic ash, and mud laid down by eruptions and pyroclastic flows, all of which nurtures fields of strawberries, blueberries, sweet corn, and wheat. The summits of the San Juan Islands to the west remind me of reptilian beasts—Lummi, with its dragon's-back spine, Orcas, with its three-thousand-foot summit and turtle's back, Cypress, a slumbering brontosaurus.

A summer rain squall precedes us. The campground sits back from the bay in a stand of second-growth Douglas firs. The "wet camping" sites, those with power and water and sewer, are full, each parked RV mounted with a satellite dish. We see no people. Maybe they're watching *Jeopardy* or posting puppy pictures on Facebook. The park ranger assigns us a space in the regular campground but advises us to go the "wrong way" around the one-way loop to ease backing into our spot. The ranger is right. It's tight. With Susan directing, I manage to steer the Bambi into the campsite. Were our trailer even a couple of feet longer, it might not fit. But it does and we are docked for the night, "dry camping" with no power or water. We don't unhitch—we plan to leave early. I set the stabilizing jacks. Susan rewards me with a martini.

From my camp chair, I contemplate the nature of Bambi travel. I'm reminded of a Chinese tradition I learned about when we lived in Hong Kong. As a young person, you follow Confucian ethics: family, mentors, community, nation, duty. As you age, you can opt for a different spirituality, Buddhism, or Dao. Dao explores the connection between natural things. This day, Dao seems perfect. Connecting the present and the past. Retrospective travel.

Farewell Bend and Bay View aren't the beginning of our Bambi travels, nor the end, but places to begin the diary. Perhaps it is not so much a diary as a travelogue, hopping around in time, combining different trips, uncoupled from the tyranny of calendars or a specific journey.

THE SUBDUCTION ZONE

US Highway 101

> Let me recommend the best medicine in the world: a long journey, at a
> mild season, through a pleasant country, in easy stages.
>
> —JAMES MADISON

WE'RE CAMPED ON THE Washington coast at Kalaloch Campground in Olympic National Park, adjacent to US Highway 101. Our site lies a hundred feet from a clay cliff that drops down to a long gray sand beach, laced with foam. Shore pines, cedars, and Douglas firs, storm-bent, arch over our campsite. A light rain drizzles. We're seated in the dinette of our Airstream Bambi trailer sipping glasses of Washington State chardonnay. We are celebrating our Bambi purchase. From the dealer this morning, we headed straight for Kalaloch Campground. For me, no other place would do. Kalaloch is an archeological dig from my camping boyhood, layer over layer of trailering memories. Here my dad and my brother Charlie and I camped the 4th of July weekend in 1956 after my sister was born. Here Charlie and I, on many weekends, unsuccessfully surf-fished. Here we lay awake in our sleeping bags listening to the Destruction Island foghorn, now stilled. Here my brothers and sister and I leapt from one silver drift log to another, the logs hundreds of feet long and scattered like jack-straws across the upper beach. Here I stood on the beach and tried to imagine what it would be like to cross the Pacific, to visit Japan and Hong Kong and Singapore. Here my grandmother last camped. Here my sister spread our grandmother's ashes.

To get to Kalaloch from Seattle you travel south on Interstate 5 through Tacoma, turn west at Olympia on US 101, follow State Route 8 through Elma and Satsop (under the shadow of the cooling towers of the never-to-be-built Washington State Public Power System nuclear plant, a.k.a. "Whoops" for its ill-conceived but premonitory "WSPPS" acronym), continue past Montesano, head north at the down-on-their-luck mill and fishing towns of Aberdeen and Hoquiam, again on US 101, travel inland through logged-over and replanted clear cuts, transit the not-really-even-a-town of Humptulips (a great name nonetheless), enter the Quinault Indian Reservation, pass through Amanda Park and Queets, and then, after a dozen miles, enter Olympic National Park. In a few minutes, you'll arrive at Kalaloch Beach. There's a small concessionaire's lodge, a store, and a covey of lodge cabins. Kalaloch Creek flows under the highway bridge and enters the ocean below the lodge. The beach is flat, its sand gray, its berm silvered with driftwood. Low upland cliffs back the beach, cloaked in western red cedar, Douglas fir, and shore pine. In the Quinault language Kalaloch means "a good place to land a canoe." The Quinault must have been good boatmen—in all seasons, waves break on this beach in white-foam fury.

My family came here the weekend of the 1964 Good Friday Alaska earthquake, a 9.2 Richter whopper, the largest recorded quake in North American history. The day after, rangers drove through the campground warning us to be ready to leave in a hurry—that there might be an aftershock-triggered tidal wave. One hundred and forty-seven people died in that quake, most in Alaska. What we didn't know then was that four had died south of us on the Oregon coast, at Beverly Beach State Park, children swept to sea in the quake's initial tsunami. We saw no tsunami. We didn't expect one, either, an expectation born of ignorance. Only a few places in North

America can trigger a quake of 9-plus magnitude. One is right here, off this beach, where the Juan de Fuca Plate dives under the North American continent—"subducts" is the technical term. This is not a smooth process. Subducting occurs in fits and starts, what geologists call "fault rupture." Pressure builds up. Pressure suddenly releases. Geophysicists believe a major quake occurs on the Juan de Fuca Plate once every five hundred years. The last one was in 1700. Although no one in North America left a written record of it, we know the date with unexpected accuracy because its tsunami arrived in Japan on the evening of January 26, 1700. We are three hundred years into the cycle. The specter of this quake—a certainty—hangs over Seattle, Vancouver, and Portland, the Northwest's skyscraper metropolises.

Can ignorance be a coping strategy?

Yes, albeit not a good one.

Susan and I sit at our dinette and look out the Bambi's window. We tell each other that this beats the heck out of tent camping. Despite the drizzle, an elderly couple is walking arm in arm through the campground. The man is walking with some difficulty, the woman supporting him. I find something beautiful in this couple, here at Kalaloch, on this rainy coast, in the sunset of their lives.

Later Susan and I take a long walk down the beach. Gulls strut at the water's edge. The surf ebbs and flows around their feet. The break is modest, although it still breaks white. I pick up a stone. It's the size of a Ping-Pong ball and pocked like an asteroid. This stone was birthed on the ocean's floor in a rain of river-outflow sediment. It's called mudstone and is a form of chalcedony, an ancient name for a type of quartz, the most common mineral in the continental crust. I think of my grandmother's ashes spread on this, her favorite beach. Perhaps some part of her will become mudstone too, to be found on a future beach by a future beachcomber like me.

We return to our campsite and the Bambi. It nests below a gray sky and black-green trees, silver, bright, otherworldly, seemingly beamed down from an orbiting starship.

The campsites above the beach are full: no tents, all RVs, most box-shaped and larger—much larger—than Bambi. In the last twenty years, recreational vehicles have bloated in size, what we expect of most things in contemporary America. A generation ago twenty feet was lengthy. Today thirty to forty is normal. The largest RVs, Class As, are as big as Greyhound buses, are built on bus chassis, have bus diesel engines, and feature interiors far grander than any Greyhound. Custom versions sell for a million dollars or more: Rock stars use them to reach remote venues and stock-option millionaires drive them to homecoming football games. If the RV is a trailer instead of a bus, its tow vehicle is almost always a half-ton or larger pickup. The largest trailers—some over sixty feet long—actually require a semi-tractor truck to tow them. Kalaloch is not "big-rig" friendly. Still, our fellow RVers have managed to wedge their rigs into the relatively small campsites. Older and smaller RVs haven't vanished: There's a nostalgia-driven trend toward smaller, un-boxy 1950s-style travel trailers, especially among young families. But this is off-season: The kids are in school; the weather is cool; the campgrounds are filled with retirees.

With Bambi, however, Susan and I have opted for a simple aesthetic—Zen trailering.

Our plan for this maiden voyage is to follow the Olympic Peninsula to Port Townsend, then cross Puget Sound on a ferry to Whidbey Island. Until Port Townsend, we'll take 101, a highway that will become one of our most frequent Bambi travel routes. We like it because it avoids the heavy truck traffic on Interstate 5. It passes through pleasant shoreline towns: Ilwaco and Long Beach in Washington, Cannon Beach, Newport, and Bandon in Oregon, Crescent City and Arcata in California. We like the parks: Cape Disappointment in Washington, Oregon Dunes in Oregon, Del Norte Redwoods in California. But mostly we like watching the wave-fueled fury of the Pacific Ocean crashing against the up-thrusting North American continent.[1]

US 101, a.k.a. the Pacific Coast Highway, begins in Los Angeles, touches the Pacific Ocean at Ventura, runs north through Santa Barbara, and slants inland after Goleta, where it remains until the Golden Gate Bridge in San Francisco. In Marin County, it trends east to the Napa and Sonoma Valleys and remains inland through the redwoods country until it reaches the northern California town of Eureka. From there it follows the coast, never more than a few miles from the Pacific. At its northern denouement in Washington State, it loops around the Olympic Peninsula like a shepherd's crook, tracks south down Hood Canal, the ice-sculpted fiord that demarks the Olympic Peninsula from Puget Sound, and terminates at the state capital, Olympia. I call the Oregon–Washington portion of Highway 101 the "Subduction Zone"—a terrain of sea stacks, pinnacles, wave-blasted capes, wind-whipped headlands, conifer-clad mountain ranges, long sandy beaches, a major continent-draining river—the Columbia—and several estuarine bays that promise refuge for mariners willing to navigate their tumultuous and dangerous river bars. This grand-opera geography owes its origin to one major phenomenon: the subduction of ocean plates under the North American continent. This makes it one of the most tectonically active regions in the continental United States and thus the area most prone to city-killer earthquakes and coast-drowning tsunamis. But for Susan and me, like most of our West Coast neighbors, its beauty makes its danger a risk we're willing to bear.

Today we're not thinking about quakes or tsunamis. With the Bambi in tow, we continue north on US 101. We stop briefly at Ruby Beach, where black basalt pinnacles rise from a gray shingle beach. To the south is Destruction Island, one of the few offshore islands on Washington's coast. Why so few? Why so (relatively) few headlands? Wave refraction is the answer. Waves approach the coast, curl around the island/headland tips, shorten their period, focus energy on the tip, pound pulse after pulse of compressed air into fissures in the rock, and crack it. Little by little, the headlands and islands wash away. In 1775, a Spanish shore party landed near here and was massacred by local Indians, an event reprised twelve years later when

Captain Charles William Barkley landed his English shore party to the same end. Barkley named the river where his crew was massacred "Destruction," a name transferred to the island by Captain George Vancouver when he restored the river to its Native American name, "Hoh." Congress failed to fund a lighthouse here until 1889, despite this being a notoriously dangerous coast. Even as it was being built, the sailing bark *Cassandra Adams* ran aground on the island's north reef. This is a foggy coast as well as a stormy one. I remember watching the lighthouse flash and its foghorn sound as I lay in my sleeping bag in Kalaloch Campground in the 1960s. The lighthouse's crew is gone now. So is its light. So is its horn.

At Ruby Beach, US 101 heads inland, not to approach the coast again until Port Angeles, a drive of some eighty miles. From here to Camp Flaherty there are only a few Native American coastal communities—Quillayute, La Push, Ozette, Neah Bay—and only a few roads from US 101 to the shore.

One of my favorite short stories concludes at Cape Flaherty. The story is by Charles D'Ambrosio. It's called "Her Real Name." The narrator, Jones, recently out of the Navy, drives west from Virginia until he meets a young woman pumping gas at a crossroads town in southern Illinois. The girl—D'Ambrosio never names her—agrees to accompany Jones on his journey. Later she confides that her stepfather, a violent fundamentalist Christian, is certainly pursuing them. Later still, Jones discovers that the girl has late-stage terminal cancer. By the time they reach Neah Bay the girl has died and a forest fire is burning. "White shacks lined either side of the street," D'Ambrosio writes of the Makah Indian Reservation, "staggering forward on legs of leaning cinder block, and a few barefoot children played in the dirt yards, chasing dust devils. Several girls in dresses as sheer and delicate as cobwebs stood shielding their eyes and staring at the fire." Jones steals a small outboard boat, motors out and around Cape Flaherty, beaches the boat, loads the girl's body into its bow, and heads seaward again. When he is beyond what he believes to be the continental shelf, he attaches a flashlight to the girl's bagged body and pitches the bag overboard. "Down she swirled, a trail of light spinning through a sea that showed green in the weakening beam and then went black."

In the Pacific Northwest, despite the subduction zone, much of our geology has nothing to do with plate tectonics. As Susan and I approach the Juan de Fuca Strait on US 101, we're entering a domain not of plates but of ice. We've camped on it at Dungeness Recreation Area. The campground is at the base of the spit, a six-mile-long run of sand, cobbles, and driftwood that cuts into the strait like a scythe blade. We've hiked halfway down the spit. It's a cool, clear day in May. The strait is calm, the weather dry. Farmers irrigate here on this corner of the Olympic Peninsula because the mountains dry out the Pacific storms, creating what is called a "rain shadow." To the north, we see Victoria on Vancouver Island, the capital city of the Canadian province of British Columbia. To the west is Port Angeles, the Clallam County seat, to the east, the clay banks of Whidbey Island, the fourth-longest island in the contiguous United States.

I pick up a stone. What clues does it offer about place? The stone is polished and smooth and red as the planet Mars. The stone is jasper, another member of the quartz family, and it forms when fine-grained sediments—mudstone, sandstone, or volcanic ash—cement with silica in a process called diagenesis, which may be chemical, physical, or even biological, but which takes place at temperatures not so hot as to melt the rock, not so exposed as to weather it. The jasper's smooth surface is due to its original fine-grained texture being enhanced by waves, which raise up these stones and roll them together. Waves are the genesis of Dungeness Spit. Driven by winter westerlies and by summer northeasterlies, waves lengthen the spit by sixteen feet each year. It's unlikely the jasper was born here. This stone traveled a great distance, carried by the continental ice sheet that, during the last ice age, filled Juan de Fuca Strait. Even in high Olympic Mountain valleys, you will find stones that have their origin in the far north of British Columbia. This is a geography of ice, with glacial kettles and glacial drumlins and solitary glacial esker rocks that can be as big as houses.[2] Mammoth tusks have been found here, one as long as sixteen feet, in the glacial till of this same Dungeness Recreation Area. These are estimated to be a hundred thousand years old but mammoths may have grazed here much more recently, perhaps only four thousand years ago, an evolutionary blink of the eye.

Our campsite is adjacent to a group site where on this May evening a dozen or more children run back and forth in a game of Capture the Flag. Several hide under the bushes in our camp until a young girl, perhaps twelve years old, her hair as golden as the sun, stops and apologizes to us for disturbing our evening peace. But we are not disturbed. We are nostalgic. The games and laughter remind us of our son John and our nephews and nieces, moved on to other games, too old now for Capture the Flag. Why does it seem that we heard their laughter only yesterday? A tick in time, a blink of the eye.

At the Port Townsend Ferry Terminal, the attendant gives us a "thumbs up" as she directs us onto the ferry. Later, at the Anacortes Ferry Terminal, another passenger knocks on our car window and asks to see inside the Bambi. We will become used to this: people smiling at Bambi as we roll by, highway workers shouting out approval, strangers peering into our open door. The Bambi sparks interest, gives pleasure, evokes a spirit of travel. It is an icon of the road.

Wally Bynum, the Airstream founder, wanted a trailer that would "flow like the air," although Bynum didn't actually design the "silver bullet" model for which the brand is famous. That was Hawley Bowlus, better known for designing Charles Lindbergh's *Spirit of St. Louis*, the airplane that flew the first nonstop flight across the Atlantic. When Bowlus's company began to fail, Bynum bought him out and launched the 1936 Airstream Clipper, essentially Bowlus's design with the door moved to the side. The Clipper epitomized the late Art Deco style known as "Streamline" or "Art Moderne," which featured horizontal orientation, rounded edges, smooth exterior surfaces, subdued colors, and lots of aluminum. (Other Art Moderne designs included the New York Central steam locomotive the 20th Century Limited, the Puget Sound ferry *MV Kalakala*, the Star Ferry Pier in Hong Kong, the 1939 New York World's Fair buildings, and the Emerald City set in the movie *The Wizard of Oz*.) By the 1950s and 1960s, Art Moderne had become retro. "Ray-gun Gothic," science-fiction writer William Gibson called it. He went on to name an alternate-universe future in one of his stories "The Airstream Futuropolis: the Tomorrow That Never Was."

But if you asked most people what travel trailer design most epitomizes the idea of the open road today, I suspect many would cite the Airstream.

Is functionality what makes the Airstream iconic? The company describes its design as "functional" but the Airstream shape is not the most efficient shape for interior space. There's a reason why most travel trailers are boxy, and plastic and composites are now lighter than aluminum. My San Diego brother's trailer is two feet longer than our Bambi and weighs five hundred pounds less. Even the Airstream's pod-like shape has drawbacks. A ticket-taker at the Anacortes Ferry Terminal once told me she'd lived in an Airstream and found that because of its curved architecture, "there's no place to hang pictures."

Functionality is not the answer. Style is. As with any good icon, its form conveys its purpose. The "silver bullet" conveys movement. Movement conveys travel. Travel conveys freedom. And these attributes make the Airstream iconic, so much so that film stars write Airstreams into their contracts as movie-set refuges, museums feature them in design exhibits, and restaurateurs modify them to serve as upscale food vans.[3] "It's a chick magnet," one Airstreamer cautions Susan.

Several months after our inaugural Bambi journey, we're in Oregon on an Indian summer Sunday. We're making a run south down the full length of Highway 101, all the way to Los Angeles, but we've detoured into the Willamette Valley in order to taste a little wine and to effect a minor repair to the Bambi. We're camped at the junction of the Willamette and Clackamas Rivers in a city-owned RV park. The place is light on amenities—just a picnic table, water, and electricity, the restrooms a half mile distant in the picnic area—but it is close to the Airstream dealer and the setting is lovely: a flat, grassy field adjacent to a cobbled river beach. Cottonwood trees overhang the field. A half-dozen carnival rides glitter behind the park. Across the river, suburban homes top a high-bank shore, each with a dock fastened to twenty-foot-tall pilings, their height necessary to deal with fluctuations in the river's level. The water is low now—it's September—and the beach is wide. A few cars and trucks have driven onto the cobbles. A man and a young woman are hauling out video equipment from a large SUV. Soon the man begins filming the woman dancing with hula-hoops. She's very good. Two, three, four hoops at a time. A whole stack of hoops. The sun is setting. The air is cooling. Susan and I retire to the Bambi. Midway through watching a TV program we notice flickering light in the darkness outside. We draw the rear-window curtain open. The hula-hoop woman is dancing just yards away, her hoop aflame, whirling around her waist, the videographer filming her, her fiery dance mirrored in the Bambi's aluminum.

Chick magnet.

CHAPTER 3

AMERICA'S ROOFTOP

US 550, Colorado SR 189, and the Million-Dollar Highway

> A good traveler is one who does not know where he is going to, and a
> perfect traveler does not know where he came from.
>
> —LIN YUTANG

IT'S JUNE. WE'RE IN Durango in southwestern Colorado, at 6,500 feet. To get here, we motored over Washington State's Cascade Mountains, crossed the Columbia River at Umatilla, Oregon, cut across the northeastern corner of that state, drove to Boise and then across southeastern Idaho, entered Utah, passed north and east of the Great Salt Lake, then climbed the Wasatch Mountains and headed across the midriff of Utah to Moab (unofficial motto: "Mining, Mormons, Mountain Bikes"). In Moab we spent a few days mountain biking and hiking in Arches and Canyonlands National Parks, then a couple of days exploring Ancestral Pueblo cliff dwellings in southwestern Colorado's Mesa Verde National Park. All the way, we've been gaining altitude. Now we're poised to transit north through western Colorado, skirting the Continental Divide, America's rooftop.

This will be our third Bambi entanglement with the Continental Divide. Two years ago we followed its eastern slope from Canada down into Montana and Idaho. Last fall we crossed back and forth over the divide while visiting Yellowstone, the Grand Tetons, and Wyoming's Wind River Mountains. The term "continental divide" is one of the least capricious in the geological lexicon because it's determined by answering a simple question: To which ocean do the local streams and rivers flow? Following the divide, however, can seem capricious. It zigzags back and forth, usually tracking the mountain ranges that divide the continent, but not always. Sometimes it goes east–west rather than north–south. Sometimes it loops back on itself. Sometimes rivers and streams flow to no ocean at all, as in the Great Basin of Nevada and Utah and portions of Wyoming south and east of the Wind River Mountains. On this trip, in Colorado, we'll be mostly on the divide's western slope.

For sons and daughters of the West Coast like Susan and me, Colorado marks the eastern boundary of the West, although only the western third of the state (in our opinion) is truly "the West." The eastern two-thirds are pretty much flat: no great mountain ranges, not exactly Kansas, but not Utah or Nevada or California either, even though Wallace Stegner, in his book *Beyond the Hundredth Meridian*, maintains that the West begins at the twenty-inch annual rainfall line, which more or less corresponds to the 100th meridian, which would include all of Colorado. But how can anyone call it the "West" when it has no mountains? Call the eastern two-thirds "high plains" if you want, but not "West."

Colorado is not only closer geographically to the Midwest and South but closer culturally. In Colorado, you begin to hear the slower-paced accents of Texas and Oklahoma. You begin to see license plates from further afield than we're used to: Kansas, Nebraska, and Wisconsin; Minnesota, Illinois, and Iowa; Oklahoma, Arkansas, and Texas. In the Alpenrose RV Park, where we're staying now, these relaxed accents prevail. Many visitors here—especially the Texans and Oklahomans—have arrived for a summer-long stay, escaping, we would guess, the heat of their home states.

Durango is pleasant and we like it despite its touristy identity. It was founded as a railway town in 1881 to support the San Juan Mountains' gold and silver mines, which ultimately played out. Tourist dollars replaced gold and silver: winter skiing, summer off-roading, fly-fishing, mountain biking, hiking, climbing, and rafting. The Durango and Silverton Narrow Gauge Railroad is a big deal tourist-wise. Its steam engines haul tons of visitors forty-two miles and three thousand feet up to the old mining town of Silverton. Susan and I forgo the Durango and Silverton, opting instead for the paved bike path that follows the Animas River from one end of town to the other. We cycle its full length, fourteen miles round trip. When we stop for a water break, we watch rafters drift downriver and through town.

The RV park is nearly full, only a few spaces available. We're only beginning to understand the RV community's sociological and cultural fault lines. Some are demographic: baby boom retirees versus Greatest Generation retirees versus Gen-X/millennials with their families. Some are economic: big-rig versus small-rig; neo-trailer-park versus nearly homeless. Some are lifestyle choices: vacation travelers (who move every few days) versus "Snowbirds" (who migrate each season to more hospitable climes), "Full-Timers" (who have abandoned real estate for a mobile residence), and "Workampers" (who follow jobs while living in RVs), all of whom more or less stay put. For some, RVs are a means to a truer passion—hunters, fly fishermen, ATVers, snowmobilers, surfers, football-game tailgaters, swap-meet attendees, square-dancers, Civil War reenactors, collectors of practically anything, festival-goers (Deadheads, Burning Man, Christian music), Seventh Day Adventists, and many, many others. There are "RV restorers" who are passionate about a specific brand (one of the Airstream subcultures). There are RVers who join groups like the Airstream Wally Bynum International Caravan Club. Some individualist RVers forgo social connections and seek solely the solitude of the road. Here in Alpenrose RV Park, we have a cross-section.

The tent-trailer guy next to us—a baby-boomer retiree—hails from Missouri. He tells me he's on his way to Oregon to fish for salmon. "Moved the boat earlier," he says. His wife plans to fly to their destination because "she doesn't like to drive."

The guy on the other side of us is solo: gray-haired, gaunt, and silent. He has a thirty-foot trailer, a black pickup truck, and a morose-looking German shepherd. He spends his time "clicking in" the sights of various rifles using a sort of sight-calibration apparatus he has set up in the bed of his truck. Susan and I find the rifles unnerving.

A seventy-something woman stops by and introduces herself as the president of the Washington State Chapter of the Wally Bynum International Caravan Club. She tells us that she has been traveling alone with an Airstream for a dozen years. Her present model is an older Class B Airstream, meaning not a trailer but a truck-based model. She's on her way to Farmington, New Mexico for the Western Regional Caravan Club Rendezvous. "Are you heading there too?" she wants to know. When we tell her no, she's politely disappointed.

Two sixtyish women with Texas accents knock on our screen door one morning before I've had my first cup of coffee. They want us to know that our trailer is "cute." This in itself is a little unusual. Serious RVers don't regard the Bambi as a real RV. Or maybe "cute" means "not real."

A number of the older retirement-age guests seem to know each other. They're regulars, we guess, something we've seen elsewhere, especially in scenic and recreational destinations like Durango. Retirees definitely dominate Alpenrose, although maybe that's because it's early June and most families have kids in school.

A fly fisherman stops by and explains that he's thinking about heading elsewhere because the snowy and rainy Colorado spring and a heavier-than-usual runoff have made the streams turbid and the fishing poor.

A male Boomer retiree surveys the Bambi and our fully loaded Acura MDX with its inflatable kayaks and paddle boards, bike racks, and bikes topside and asks somewhat skeptically, "That car tow well enough?"

"Well enough," we assure him.

The highway that runs in front of the Alpenrose RV Park, US 550, is also known as the San Juan Scenic Skyway. From Durango north to Silverton it follows the Animas River and loosely parallels the Durango and Silverton Narrow Gauge Railroad. The next twenty-five miles, from Silverton to Ouray, are called the "The Million Dollar Highway." There's some disagreement about the name's origin. Some say it was the cost to build the highway in the 1920s, others that it's because the highway's fill dirt contains a million dollars of gold ore. What's not in dispute is that this is one hell of a mountain highway: three passes, Red Mountain, Molas, and Coal Bank, each over ten thousand feet (Red Mountain is over eleven thousand), hairpin curves, narrow lanes, no guardrails, steep cliffs (a website warns about "the vertiginous outside edge of the highway"), and three thirteen-thousand-foot mountains—Sultan, Kendall, and Storm Peak—looming over it all. The highway is open year-round. Twenty-five years ago, Susan, our son John, John's nanny Vilma, and I drove it southbound during a winter ski trip. This time Susan and I will be heading northbound. We'll have less snow to contend with but will be towing Bambi.

I don't find driving mountain passes particularly intimidating, I suppose because my father drove over so many when I was a kid that what fear I had was scared out of me. Trailer towing, however, warrants a few points of extra consideration. First is the power of your vehicle. You need enough power to keep climbing at a respectable speed or you'll end up like Lucille Ball and Desi Arnaz in the 1953 movie *The Long, Long Trailer*, with a line of horn-tooting, angry drivers queued up behind you. The second issue is gravity. When descending, gravity is trying to hurl you down the mountain at a speed faster than it's safe to go. There are two ways to slow down: One is applying your brakes, the other, using your tow-vehicle engine torque as a brake. Most trailers have electric brakes that operate in conjunction with the tow vehicle's brakes, either automatically (you can adjust when they kick in) or manually, using a lever on the brake controller, which, in our car, is mounted under the steering wheel. I never feel comfortable groping for the brake lever, especially when I'm steering around hairpin turns. My preferred method is engine torque with modest application of the brakes. You put the transmission in a lower gear so the engine acts as a brake, essentially pushing back on the gravitational force that's pulling you down. But there's a caveat: You need an engine powerful enough to generate the torque to offset the weight of your tow vehicle, passengers, trailer, and all the bikes, kayaks, barbeques, and whatever else you've dragged along. You also need an automatic transmission that will remain in a lower gear—some won't. The other major concern on mountain passes is other RVs, especially big rigs from flat states like Kansas, Florida, and Illinois, whose drivers tend to hog the middle of the road and may veer around a hairpin turn, tensed in white-knuckle horror. Still, given all this, I'm confident in the MDX and the Bambi. We've summited several nine-thousand-foot passes before, although Red Mountain will be our first eleven-thousand-footer.

We've picked a good day for the Skyway—sun out, not too hot, traffic light. Colorado's late snowfall and wet spring may have hurt fly fishing but it's heightened the scenery. The mountain summits remain snow covered. Cottonwoods and ranchland grasses green the Animas River Valley. The red sandstone and dark pines of the valley-side Hermosa Cliffs complement the valley greens. We climb gradually at first, but eventually the road begins to spiral upward. We summit Coal Bank Pass, then Molas Pass, both ten-thousand-footers. From here we can see Molas Lake, the Animas River Gorge, and Snowdon Peak. We haven't reached the really narrow-lane segment— that comes later, after Silverton. We see tailings from old gold mines in several places. Mines scar the West, but perhaps more so in the Rocky Mountain West and, it seems to me, most of all in Colorado. This may be because the roads we drive were originally mining roads, or because Colorado was one of the first Rocky Mountain states to be mined. Maybe it's because the geology of the Coloradan mountains makes the most valuable minerals accessible.

Most people oversimplify mining—you're either for it or against it. It's not that simple. Colorado's mines (and mines elsewhere in the West) have fueled American industry and raised the standard of living for most of us, as well as for our parents, grandparents, and great-grandparents. Colorado mines, with their silver and gold, financed the defeat of slavery in the American Civil War and provided minerals necessary for victories over Nazi Germany and Imperial Japan in World War II and Soviet communism in the Cold War. Mines continue to contribute to American economic strength: Much of what we take for granted in our twenty-first-century lives depends on mined minerals. At the same time, mines, especially those that have been abandoned, can be environmental time bombs. Susan and I don't know this at this point in our trip, but in two months one of those time bombs, one very close to where we are now, will explode.

Here's what will happen:

On August 5, 2015, Environmental Protection Agency personnel and EPA subcontractors using a backhoe will inadvertently displace a plug holding water trapped underground in the Gold King Mine, abandoned in 1923. Three million gallons of mine waste and tailings, including cadmium, lead, arsenic, beryllium, zinc, iron, and copper, will dump into Cement Creek, a tributary of the Animas River. The Animas will turn yellow as dissolved iron from the spill oxidizes, photos of which will be broadcast around the world. The contaminated water will flow into the watersheds of Colorado, New Mexico, Utah, and the Navajo Nation. It will be revealed that this disaster occurred because: 1) the EPA personnel attempting to install a drain in the mine tailing pond in order to prevent exactly this kind of event did not know that the mine was already flooded; 2) the agency was forced to cobble together this solution because a more permanent and more expensive one—designating the region and its abandoned mines as a Superfund site—had been blocked by local authorities who feared their tourism-based economy would be tainted by the designation; and 3) some EPA personnel suspected that the Gold King Mine might flood because it had shared drainage with the two other mines, the Sunnyside (closed in 1991) and the Mogul (sealed in 2003), but did not inform all of the EPA players. Ironically, the attempts to "seal" the other mines inadvertently contributed to the Gold King flooding.

Is anything simple here?

Not much. Not the mines' interconnectedness. Not the expensive long-term solution. Not the conflicting governmental jurisdictions. Not the need to cover your ass if you're an EPA employee or contractor. Not the jobs that might be lost in the tourist industry.

And how clean are *your* hands? If you turn on your electric lights, use a cell phone, wear a gold wedding ring, watch TV, drive a car, undergo a CAT scan or a root canal, or get an X-ray, your hands are not clean. Heavy metals are essential to all these technologies. And even if the metals didn't come from Colorado, they came from deep below the earth somewhere.

Susan and I, however, on this bright, cool June morning in 2015, know nothing about the Gold King Mine disaster because it hasn't happened yet.

Silverton, at 9,300 feet elevation, doesn't feel like June. There's a chill breeze even though the lower mountain slopes are snow-free. Grass greens the town lawns, the aspens bud spring green with new leaves, and the small wooden houses display flowers on their windowsills. Still, something raw and dusty characterizes the place. The main road, Greene Street, is paved, but not back streets such as Reese Street, where we find room to park the car and Bambi. The town hops with tourists. Most arrived via the Durango train. But a surprising number came on their own. Off-road vehicles—jeeps with electric winches mounted on their front bumpers, jerry cans on their tails, and shovels strapped to their sides—jam Greene Street's parking spaces. "Jeeping" is big here. The mountains are laced with old mining roads on what is called the "Alpine Loop," sixty-five miles of unpaved tracks between Silverton, Ouray, and Lake City, which include seven ghost towns, two eleven-thousand-foot passes, and numerous primitive campsites. At lunch, we wait for a table. The rail-way tourists arrived before us. But they'll eat fast: they have only two hours in town before the train whistle summons them back. After lunch, we explore. Tourist-treasure shops. A few more restaurants and bars. We find the US Post Office. We don't find the "Our Lady of the Mines" shrine.

Beyond Silverton, we ascend Red Mountain Pass. We're feeling the altitude, not for lack of oxygen but for how it looks: The trees have thinned out and the sky has turned a deep, high-altitude blue, almost as if we could reach out and touch an infinity of color. The MDX and Bambi are doing fine despite the eleven thousand feet. We descend in hairpin switchbacks. The "narrow lanes" and "vertiginous outside edge of the highway" that we'd been warned about seem less threatening than expected, perhaps because it's a weekday and traffic is light. The town of Ouray begins at the last hairpin turn. Ouray lies at the base of a narrow, cup-shaped valley with high mountains on three sides. A sign proclaims it to be "The Switzerland of America" but it owes no apologies to Switzerland. With its red-rock mountains and late-nineteenth-century brick buildings Ouray looks more like the perfect Western movie town, which it is. Several movies have been filmed here, the most famous being *True Grit*, with John Wayne as Rooster Cogburn, the one-eyed, pot-bellied lawman hired by Kim Darby's character, Mattie Ross, to avenge the death of her father. Wayne won an Academy Award for his role. We see no movie stars today. We do see a lot of motorcycles and an alarming number of kitschy Western apparel and tourist trinket stores. We have a premonition of what the place might look like in July, brimful of Illinois and Missouri license plates. Ouray had been a possible campground destination for tonight but it's early afternoon and we're a little put off by the tourists. We decide to continue on.

Our new destination is Ridgway State Park. We know nothing about Ridgway except its location. We plan a day trip south to Telluride on Colorado 62 (and 145) and the park is near the highway junction. After two Bambi seasons, we've learned that state parks, at least in the West, usually offer the best RV accommodations—roomy sites, clean restrooms, often power and water, dump stations, and scenic views (a state park usually is there because there's something there to see). But a paradox exists. After years pre-retirement waiting for weekends so that we could stay in public campgrounds, weekends have become the worst time to camp there. Weekenders descend on the popular parks. Almost all the park systems have implemented reservation systems to manage the demand. You can play this system, and many retirees do, if you know the specific day when park reservations open. But for Susan and me, making reservations months in advance seems counter to the spirit of serendipitous travel. Today, however, is a weekday, before the peak vacation season. We don't anticipate a problem. What we don't know is that the Telluride Bluegrass Festival begins the day after tomorrow.

On our arrival, signs are already posted that the park will be full for the forthcoming weekend. But what does "weekend" mean? Does Friday count? We want to stay through Friday.

"Shouldn't be a problem," the female ranger at the park gate assures us. "Just find a site not reserved for Friday."

And we do. A glorious site.

Susan and I (and maybe most campers) have this idiosyncrasy: We'll explore every campground loop to find the best site, even though we won't be in any site for more than a couple of days. "Best" is based on a set of attributes: a level parking place for Bambi; ease getting in and out, especially if backing; a degree of privacy from adjacent sites; not too close to or too far from the restrooms; shade if a hot climate; good neighbors (avoid high-school groups, crying babies, campers brandishing weapons); no nearby RVs that look as if they'll run their gasoline generators late into the night; no neighboring campers who are already intoxicated—it isn't going to get better; a windbreak if windy; proximity to any creeks, rivers, lakes, or oceans; and a view—this last being an intangible that may include mountains, bodies of water, picturesque trees or cacti, interesting geological formations, or fields of wildflowers.

Our Ridgway site scores on all counts: level and paved, a pull-through (meaning no backing), no neighbors, restrooms two sites distant, electricity, water within hose reach, a shade shelter and modest shade tree, and an open grassy slope within view of a reservoir lake and two snowcapped mountain ranges, one east and one south.

You can't do much better than this.

After we settle in (Bambi unhitched, stabilizing jacks down, electrical power connected, cocktails prepared and served), I haul out my *Roadside Geology of Colorado*.

This particular volume is coauthored by two women, Felicie Williams and Halka Chronic. A separate book exists for each Western state. I've brought the books for the states we plan to visit. Chapters are arranged by highway, so it's easy to track where we've been. Today we drove US 550 from Durango through the San Juan Mountains to here. And today, from the book, I find a surprise. What I expected today was layered sedimentary rock, the signature rock of the Southwest, what you see in Zion National Park, the Grand Canyon, and Monument Valley, albeit sculpted by Pleistocene glaciers here in the high-elevation San Juan Mountains. Indeed, we did see sedimentary formations around Ouray and in the Animas River Valley. What I didn't expect were volcanoes. According to Chronic and Williams, the geology from Silverton to Ridgway is largely volcanic due to large, explosive calderas that erupted twenty to thirty million years ago. This was followed by fault-driven basalt flows that covered wide areas of western Colorado.

This child of the Pacific Northwest expects volcanoes to look like Mount Rainier or Mount St. Helens—cone-shaped stratovolcanoes—or at least like the big slab-sided shield volcanoes Mauna Kea and Mauna Loa in Hawaii, or even the black columnar cliffs of Columbia Basin flood-basalt flows. One difference here is age. The Cascade volcanoes and the Hawaiian volcanoes are young, geologically speaking. The San Juan Mountains are twenty-three million to twenty-nine million years old, the eroded and faulted remnants of nineteen enormous calderas.[1] Why this volcanism occurred in the first place is a little uncertain. Normally volcanoes erupt two to four hundred miles from a tectonic plate boundary—the Cascade volcanoes, for example—or over a "hot spot" magma plume like Yellowstone or Hawaii. But the San Juan volcanoes erupted hundreds of miles further east from where the Pacific Plate was diving under North America. California had already arrived; the plate boundary was there. The most accepted theory, according to Williams and Chronic, is that the angle of subduction by the Pacific Plate under North America was for some reason shallower, causing the volcanoes to erupt farther east. But why shallower? The authors can only speculate: "Was the subducting plate stuck beneath the Rockies?" they write, "Or was it pushing against the extremely stable area that underlies the Great Plains?" This underlies other big questions: Why are the Rocky Mountains where they are? Why do they cover so large an area? Why are they so high? Why are they still rising?[2] I love such geological puzzles, the unfolding of the land over eons, the saga of time. Some people simply enjoy looking at a mountain. Some have to climb it. I want to know why it's there, one more reason l love Bambi travel.

If serendipitous travel is one of our Bambi travel objectives, our next few days are ideal. We drive south to Telluride, intending to mountain bike around its environs. As we approach, however, reader-board signs warn of heavy traffic and parking restrictions. It turns out that the bluegrass festival has already begun. Cars are being stopped at a roadblock and directed to a large parking lot outside of town. A ski-instructor-trim, middle-aged man wearing a Telluride Bluegrass Festival T-shirt and a hat that says "Parking" waves us to a stop. "Pass?" he asks.

"What pass?" we answer.

"Your hotel's festival parking pass."

"No hotel," I reply. "Here to bike and lunch. Didn't know the festival had begun. Until yesterday we didn't even know there was a festival." I'm already calculating if it's worth the hassle of remote parking and shuttle buses.

"How long do you need?"

"Two hours."

He hands me a slip of paper. "Display this in your car window. Good for four hours. Park anywhere you can. Enjoy!" He waves us through.

Were there a contest for America's most picturesque ski town, Telluride would surely contend. With its solid nineteenth-century stone-and-wood-framed buildings and its cottage-style houses, the town would be handsome anywhere, but Telluride sits in a steep-sided amphitheater valley surrounded by snowcapped mountains, waterfalls, and wildflower meadows. We opt for an outdoor streetside lunch at the New Sheridan Hotel. Flags bedeck the main street buildings. Festival-goers wander from shop to shop—jewelers, art galleries, upscale Western and outdoor clothing. Across from us, in a small park, bluegrass bands are already performing. We linger over a bottle of wine, mountain biking forgotten, serendipitously charmed.

The next day, from Ridgway, we head north toward Montrose on US 550. We have only a loose plan. There's a National Park nearby that we've never visited—Black Canyon of the Gunnison. And east of that is a National Recreation Area called Curecanti, sited on the shores of two reservoir lakes. Beyond them is the town of Gunnison, gateway to Crested Butte, a ski resort town about which I've long been curious. Crested Butte will be our "Turn-Around Destination," meaning the place from which we'll begin heading home. We have three or four days before we plan to start back.

We arrive at the Black Canyon's South Rim Campground around noon, a good idea if it's Saturday and you don't have a reservation. The campground lies in a dry-oak forest a half mile or so from the canyon rim. One loop has electricity, not the norm for a National Park. All those sites are taken. But on the nonelectric loops, we find a pleasant site shaded with dry oaks that shield us from our neighbors. Susan organizes a late lunch, then we set out to explore the canyon.

The Black Canyon of the Gunnison was given its name because it's so deep—1,700 to 2,500 feet—and so narrow—1,300 feet rim to rim at its closest points—that for most of the day it's completely shadowed. I wouldn't call the canyon typical of Colorado geography: It lacks the sedimentary layers of the southwestern half of the state and the big, glacier-shaped mountains of the northwest, but the forces that created it are the forces that created all of western Colorado. The canyon rock is very old—Precambrian, some billion-and-a-half and more years old—and very hard because it lay for so long seven to ten miles below the surface, subjected to intense heat and pressure that folded it, bent it, and refolded it, thereby crystalizing and compounding the rock's mineral structure. It's the rock, Williams and Chronic tell us, that underlies most of Colorado, its "basement" rock.

I find it difficult to wrap my mind around 1.7 billion years and what can happen in that time. Were we to travel back to that age the continents wouldn't resemble today's continents, wouldn't be on the same place on the planet. Life, such as it was, would be simple-celled and marine. As the eons unfolded, we would watch continents drift, break apart, crash into each other, and reform, seas flood in, seas ebb away, deserts form, seas return, as each step leaves in its wake what will become limestone, sandstone, and shale in a future age. Picture these as layers in a cake. Some will be so completely eroded or washed away that they'll leave no trace. Seventy-two million years ago, six million years or so before the Cretaceous Extinction Event that kills the dinosaurs, Colorado really begins to rise in what is called the Laramide Orogeny—"orogeny" being the process of mountain formation—kicking off a tempestuous era, the Cenozoic, which will be sixty-six million years long and include volcanoes, fault rifting, further uplifting, and glaciation, ultimately creating the Rocky Mountains we know today. About two million years ago, the Gunnison River begins to flow, slow and meandering at first, gaining speed and power as the land uplifts. It erodes layers away. Volcanic activity in the Elk River Mountains diverts the river south, then southern volcanism from the San Juan Mountains turns the river north. Finally the river erodes through the Cenozoic layers and reaches the hard, ancient Precambrian rock. About this time, five million years ago, the entire region uplifts another five thousand feet to its present altitude. The river, unable to leave its valley, as Williams and Chronic put it, "scoured its way downward, pounding, hammering, and deepening a canyon scarcely wider than itself."[3]

Susan and I stand at an overlook and peer down into the canyon. In the flat, bright sunlight of midday, with the canyon so shadowed, it's hard to have a sense of its depth. We decide to visit again in the morning when the sun is just rising.

And we do.

Two days later, we're in Crested Butte. As with most other Colorado ski towns, Crested Butte began as a mining town, first for gold and silver, later for coal. There's only one RV park and it's located in a light-industrial area south of town. Simple. Clean. Open. Flowers between sites. A mix of big-rig buses and modest tent trailers. The Slate River flows behind the park. Signs have been posted explaining that short-term fishing licenses are available in the park office. The mountains around the park look more like the Rockies than what I expected: broad-shouldered, with softer summits and wide valleys. The butte for which the town and ski resort are named is a few miles east. We drive to the ski-mountain base. European-style high-rise condos crowd the ski lifts, a little out of place, it seems to us, in such a wide-open Wild West setting. The old mining town, however, charms. There are wood-frame buildings, aspens, flowers every-where. A creek tumbles beside an outdoor café. The town is smaller than Telluride and Durango and more laid-back. Outside the old town, houses, bike paths, and playfields offer a distinctly suburban aura.

Normally in Bambi travel, Susan and I bring road bikes. On this trip, because of our earlier visit to Moab, we've brought mountain bikes. Ours are "hard tails" without fancy suspensions, not what self-respecting hard-core, hard-body mountain bikers would ride. But we're amateurs and happy to stay that way. Mountain biking is more like skiing than road biking. Balance, physical strength, and courage rule the day. Crested Butte claims to have invented the sport. A network of bike trails ring the valley, a few paved, most bare dirt and single-track, meaning a hiking trail rather than a road, where riders must travel in single file. A rating system ranks each trail by difficulty, although even the brochure we picked up at the Crested Butte Visitor Center cautions that their rating system will be more challenging than elsewhere "because it's Crested Butte." Tomorrow is my sixty-eighth birthday. In the spirit of serendipity, we decide to celebrate by biking a Crested Butte trail.

Birthday day dawns cool and clear. We park adjacent to a playfield that would not be out of place in any prosperous suburb, except, of course, for the mass of Crested Butte rising in front of us. The trail ascends to the ski resort base, a well-graded, paved climb of a thousand feet. Blue lupine, golden balsamroot, and red paintbrush color the meadows. We make good progress on the paved path until we reach a steep spur road that leads to the single-track trailhead. We've chosen an "easy" trail, a reassuring green on the mountain trail map. The beginning is so steep that we walk our bikes for the first hundred yards. Then, tentatively, we mount up, wobble for a hundred yards more, find a degree of balance, and set off. We follow a ridge above green and flowered meadows. We ride through a copse of aspens. If there's a mountain biking heaven, this must be it. After a few miles, we descend a series of switchbacks, skidding perilously around each. We get lost at the bottom of the trail. Find it again along the Slate River. Follow the river until the trail widens into double track. We celebrate with lunch in a stream-side café. No broken bones. Some blood. Happy serendipitous birthday!

CHAPTER 4

FOLLOWING THE FAULT LINE

*The Pacific Coast Highway, Westport–Union
Landing to the Golden Gate*

The journey of a thousand miles begins with one step.

—LAO TZU

IT'S APRIL. SUSAN AND I are Bambi-camped in Northern California at Westport–Union Landing State Beach on the Pacific Coast Highway, officially known as California State Highway 1. Our campsite sits twenty feet from a clay cliff that drops into a churning Pacific Ocean. Sky blue, sea blue, surf breaking white. Across the highway, the Coast Range Mountains loom, dark green and blue-black, cloaked in redwoods and Douglas firs, scoured by steep ridges and deep ravines. We're sitting in camp chairs facing the ocean, our binoculars trained on a pair of Pacific gray whales spouting and diving just beyond the breaking surf. The whales are northbound on their springtime migration to the Arctic. We're southbound for a taste of California spring.

California 1, Highway 1, or the Pacific Coast Highway—all three names are commonly used—begins in Mendocino County, heart of the Redwood Empire, and terminates at Dana Point in Orange

County, home of the California beach-and-surfing culture. Except for its northernmost fifty miles, where it ascends the Coast Range Mountains from US 101 and then drops down to the ocean, it's never more than a mile or so from the Pacific. In two places, over the Golden Gate Bridge and along the coast of Santa Barbara and Ventura Counties, it's briefly subsumed by US 101. The section from Monterey south along Big Sur is famous, frequently depicted in TV commercials. It is perhaps one of the first highways built explicitly for automobile sightseeing. But the entire run is scenic and dramatically diverse: sea cliffs, pinnacles, windblown capes, sand-dune beaches. Its surfside towns might well be in New England or colonial Mexico. There are fields of strawberries, artichokes, garlic, lettuce, tomatoes. Cattle graze in ocean-side ranches. We pass by redwood groves, wind-bent cypresses, fragrant eucalyptus, oasis-like palms. Since we purchased our Bambi, we've traveled Highway 1 in sections, twice beginning at the Golden Gate Bridge and heading south. This April, we plan to follow the northern leg from Westport–Union Landing to Marin County, just north of the Golden Gate. There, our son John will join us for a weekend of Bay Area cycling and camping in Marin's Samuel P. Taylor State Park.

Westport–Union Landing State Beach falls on the simple side of the California State Park spectrum. Vault toilets. Self-sign-in. No full-time staff. The small campsites lie along an open, treeless beach cliff on what appears to be a remnant of a previous Highway 1. A weathered poster on the sign-in station bulletin board cautions abalone divers about safety. *Don't dive alone. Watch the weather. Guard against fatigue.* The poster dates from ten years ago.

The California Department of Parks and Recreation has only recently emerged from financial travail: less popular parks closed, maintenance deferred. Now most parks have reopened but many show the years of financial neglect. Here, in this simple park, it's not too bad—merely an ancient abalone-diver warning sign offering timeless advice.[1]

We plan to drive to Marin tomorrow. This is perhaps an ill-advised move given the distance—a highway this scenic deserves more time—but our son only has the weekend off. The highway is slow at any time. We encounter narrow lanes and hairpin turns, slides and washouts triggered by winter rains and fault-line earthquakes, and heavy weekend traffic as we approach the San Francisco Bay Area. We also must deal with more mundane problems: We're low on water and our dump-tank level is high. In California, because of the drought, there are few public water and dump sites.

Travel-trailer brochures and salespersons like to sell the idea of independence and the freedom on the open road. But independence and freedom are qualified. You're only independent so long as your water and propane tanks are topped off, your trailer batteries are charged, and your sewage holding tanks—the aforementioned dump tanks—aren't full. Our Bambi carries about twenty gallons of fresh water, has two five-gallon propane tanks and two deep-cycle batteries, and holds about twenty gallons of sewage or "black water." (Trailers larger than Bambi segregate waste water from sinks and showers, known as "gray water," to a separate tank even though the restrictions on dumping both are almost always the same.) Our Bambi can go three to four days without needing water or dumping waste. In California, the dump stations are mostly private and usually found in commercial RV parks. But we haven't stayed in a commercial park since we entered the state. Fortunately, websites and phone apps exist that enable you to track down dump stations. Near Fort Bragg, a few miles south of here, we've located a dump station at a convenience store. When Susan pays the bill, the clerk tells her that most people don't bother to pay. "We see them lining up," the clerk adds, "especially when we're closing the store." The dump-station bill? Five dollars.[2]

Back on the road, we soon pass Mendocino, a scenic town on a small cape above the ocean. The town's neat frame buildings resemble a New England village, a fact not lost on cinematographers, who have used it to represent that region in productions including *The Russians*

Are Coming! The Russians Are Coming!, *The Dunwich Horror*, *The Summer of '42*, and most notably the TV series *Murder, She Wrote*, where it stood in for Cabot Cove, Maine. The town has a countercultural vibe. Tourism drives the economy now (and perhaps, in the mountains east of town, also marijuana cultivation). We defer a stop today, intent on reaching Marin by early afternoon.

South of Fort Bragg, Mediterranean California begins to assert itself. The land opens. Douglas fir and redwood, though not absent, give way to coastal oaks and savanna-like hills. Cypress trees and Monterey pines crown promontories. Eucalyptus line the highway and cleanse the air with a medicinal scent. Milk-white pampas grass, red poppies, and purple ice plants rainbow the road margins.

The geology remains the same. Plate tectonics are the actor, less by direct subduction, as in Oregon and Washington, and more by lateral plate movement. As the Pacific Plate slides north, it drags the edge of the North American continent, which is moving west, creating a swarm of faults, including the infamous San Andreas Fault, fifty miles wide, whose northern terminus is under the Pacific, opposite Mendocino. Because North America is being dragged, steep valleys and deep ravines form perpendicular to the Pacific Plate path, much as wrinkles form in a beach towel when you drag one edge across the sand. This dragging isn't a smooth process—the geological term is "strike/slip." It's abrupt. Violent. From a Highway 1 driver's perspective, this means ups and downs, sharp turns, and narrow bridges spanning deep ravines. It also means earthquakes, the most destructive, at least in the recorded-history era, being the 1906 quake that leveled San Francisco. Others followed, including the 1989 Loma Prieta earthquake that knocked down Oakland's double-decked section of the Nimitz Freeway.

As Mama Cass Elliot of The Mamas & the Papas sings, "the fault line runs right through here."

We lunch in Gualala and pass through Sea Ranch, famous for a weathered-wood rustic architecture so widely imitated it has become known as the "Sea Ranch style." By the time we reach Tomales Bay, we're driving on the edge of the San Andreas Fault, which runs down the center of the bay. It's Saturday afternoon. Cars have parked a half mile on either side of the Tomales Bay Oyster Company, as if all 8.6 million San Francisco Bay Area residents have hit the road on this sun-soaked day. Motorcycle clubs. Corvette clubs. Porsche clubs. Bike racers. Kayakers. Camper vans. The joy of the Bay Area is its unparalleled beauty. The curse of the Bay Area is that it's a beauty accessible to millions of people.

Finally, at about three p.m., we roll, road-weary, into our campsite at Samuel P. Taylor State Park. We reserved this site a month ago—not our usual procedure and contrary to Lao Tzu's advice of "no fixed plans"—but we knew we'd be here on the weekend, the only time John got off work. Without a reservation there'd be no site. Ours is nestled in a grove of redwoods. Sunlight falls in slats through the trees. Campfire smoke scents the air. Kids bike up and down the campground roads. John arrives and pitches a mountain tent behind the Bambi. Our neighbors, a mother and young daughter, tent-camp next to us. The little girl, maybe four years old, wanders over to inspect the Bambi. We assure her mother that her daughter is welcome.

The key to a happy campground is following the rules. The rules are few: Don't wander into other campers' sites. Park vehicles only in your own site. Dispose of your trash when you depart. Keep the noise down, especially early in the morning and after dinner. Run your electric generator, if you have one, as little as possible and only during the specified hours.

Most campers follow the rules. The exception is the generator rule. A modern RV vacuums electricity. Air conditioning. Microwaves. Electric heat. Refrigerators. Flat-panel TVs. DVD players. Audio entertainment systems. Game consoles. Deep-cycle batteries, iPhones, tablets, laptop computers, satellite TVs, Wi-Fi routers. The largest gold-plated Class As may even have clothes washers, dryers, and dishwashers. The sounds of children's laughter, an ax splitting firewood, the campfire ballad have given way to the ubiquitous putter, clatter, and mumble of generators.[3]

On our second night at Samuel P. Taylor, a new family replaces the tent mom and daughter next door. The generator kicks on as soon as they arrive. At eleven p.m. it's still running. I dial the park ranger telephone number posted outside the restrooms. A recorded message instructs me to dial 911. Not surprisingly, 911 has higher priorities.

Discourteous neighbors aside, Samuel P. Taylor State Park is a treasure, all the more for its proximity to the heart of Marin County. Only a few miles down the road, you're on Sir Francis Drake Boulevard with its dichotomous combination of huarache-sandaled drivers in BMWs, bike shops cheek by jowl with meditation retreats, tiny nouveau-cuisine restaurants nestled against Safeway supermarkets.

We cycle the Tiburon peninsula. Watch sailboats battle the Angel Island tide. Lunch at a Sausalito marina restaurant where palm trees shade the sidewalk and bougainvillea cloaks each fence. A breeze wafts through open restaurant doors.

In Seattle, it's raining.

During the 1849 Gold Rush, tens of thousands of miners poured into the state, their kit emblazoned with the slogan "California, here we come!" Today there's no gold, especially for us. But we have received a gilded gift—a golden California springtime.

California, we're glad we came.

CHAPTER 5

YOSEMITE, MOTHER OF CALIFORNIA PARKS

> Yosemite Valley, to me, is always a sunrise, a glitter of green and golden wonder in a vast edifice of stone and space.
>
> —ANSEL ADAMS

IF THERE'S A PARK that epitomizes the virtues and the vulnerabilities of the National Parks, it must be Yosemite. If there's a park most beloved from my boyhood, it's also Yosemite. Susan and I are Bambi-camped in Yosemite's Wawona Campground, inside the southern entrance to the park, in the Mariposa "big tree" portion. Wawona lacks the geological drama of Yosemite Valley but it's lovely in its own sylvan manner. Here sequoias rise in cathedral spires. The Merced River skirts the campground, its water glass-clear, its pools deep, its banks golden with sand. We find the evenings cool, the days warm, and, at least in April, absent the frenetic activity of Yosemite Valley.

There are no vacancies in Yosemite Valley this April weekday. Visitors more farsighted (or more obsessive) than Susan and I reserved their campgrounds months ago. Soon, Wawona won't be available either, not without a months-in-advance reservation. The Park Service website offers the following advice:

Campground reservations are available in blocks of one month at a time, up to five months in advance, on the 15th of each month at 7 am Pacific time. Be aware that nearly all reservations for the months of May through September and for some other weekends are filled the first day they become available, usually within seconds or minutes after 7 am! For your best chance of getting a reservation, be sure your clock is set accurately and start the first few steps of the reservation process at www.recreation.gov before 7 am Pacific time.

I find this depressing. *Setting your clock accurately? Filled within seconds or minutes?* So much for serendipity. Of course, Yosemite's a hard case, with its proximity to California metropolises. Even in the 1960s and 1970s Yosemite summers felt crowded. Now the park's fame has spread beyond US borders. At least a quarter of our fellow campers in Wawona this mid-April hail from outside the country, mostly Europe.

Tomorrow we visit Yosemite Valley. We'll encounter tour buses loaded with Chinese, Japanese, Germans, South Americans, Russians, Italians, kids from every corner of the USA, long lines in park cafeterias, restrooms fragrant from overuse, parking lots filled to capacity. This is April. What will it be like in July? Still, the beauty of the valley overawes the crowds: Half Dome, El Capitan, Sentinel and Bridalveil Falls, the valley-wall granite that seems to glow with its own internal light. We hike across the valley among a bevy of German high-school students. We'll lunch at the Ahwahnee Lodge, this visit known as The Majestic Yosemite Lodge, cycle to Mirror Lake (which we're grateful is closed to cars), and shower at Camp Curry, now known as Half Dome Village.[1]

To get to Yosemite from the San Francisco Bay Area, you drive across the waist of California, a state long from north to south but surprisingly narrow east to west. The distance is similar to the route from Philadelphia to Washington, DC, and with similar traffic, too. Our journey began in Marin County. We traversed San Pablo Bay over the Richmond–San Rafael Bridge, where the Sacramento and San Joaquin Rivers meet the bay. These rivers sculpted San Francisco Bay during the last ice age, when the sea level was lower and the bay was a coastal valley. From the bridge, we proceeded south along the eastern shore through Richmond (oil refineries), Berkeley (the University of California), and Oakland (once home to the football Raiders, still home to the baseball A's), weaving through the remnants of the morning rush hour. We summited the Berkeley Hills and joined a convoy of semitrailers grinding toward Stockton, hangout of one of my favorite boyhood authors, Jack London.[2] From Stockton, we turned south on old US 99, now California 99, following the Central Valley, a nearly sea-level agricultural Eden (provided there's enough irrigation water, not the case in the current drought) of tomatoes, almonds, grapes, apricots, asparagus, and many other nontropical crops.[3] I have traveled Highway 99 for as long as I can remember. Once, years ago, on this stretch, my parents accidently left me at a service station, returning only when my brother Charlie noticed I was missing. I was pleased when the California Highway Patrol subsequently pulled my father over, questioning him about the whereabouts of his child.

At the town of Merced, Susan and I turned east again and began the long, slow ascent of the Sierra Nevada Mountains. The Sierras are a fault-block mountain range.[4] What this means is that the western edge is sinking down, the eastern edge tilting up as though a trap door is opening to the east. The western rise is gradual; the eastern is steep, springing from Death Valley, below sea level, to Mount Whitney, at fourteen thousand feet elevation within a few dozen miles. The Sierras rise at the westernmost edge of a geological province scientists call "basin and range," a series of valleys serrated by mountain ranges that extend from the California border east to Utah's Wasatch Mountains and south into Mexico.[5] The basin valleys and mountain ranges have formed because the earth's crust is stretching and thinning. This causes surface blocks of terrane to tilt down into valleys and up into mountains. That much is understood. What is less understood is why this stretching occurs. Perhaps it's the result of what's happening at the plate boundary between North America and the Pacific. Perhaps it's due to thermal upwelling below this segment of crust. As with so much else in California geography, it's complicated. John McPhee titled his book about the state's makeup *Assembling California* partly because so much of it originated elsewhere in island arcs similar to what we see now in the Indonesian, Japanese, and New Zealand archipelagos. The ocean plates—the Pacific Plate and the now-subsumed and vanished Farallon Plate—acted like conveyor belts, carrying these archipelagos to the edge of North America and skimming them off as the plates dove under the continent, leaving in place a multifaceted mélange. The heat generated by subduction also created plutons—blocks of melted rock that now lie like enormous granite bubbles under the earth's skin. Mountain ranges rose and the plutons thrust to the surface. Volcanoes erupted, also due to plate subduction. The result was the wide variety of rock you encounter in the Sierras. The last and perhaps the least permanent stage, Ice Age glaciation, sculpted the Sierra Nevada into its glorious valleys: Kings Canyon, Hetch Hetchy (now drowned by the San Francisco water-supply reservoir), and Yosemite, for me, its most exquisite incarnation.

From Merced, we slowly ascend the Sierras through Mariposa, Ahwahnee, Oakhurst, and Fish Camp. Central Valley fields give way to rolling oak-and-pine-covered hills, ranch country and retirement country, and, above all, Gateway to Yosemite Country. In the late afternoon, we pull into a spacious campsite in the Wawona Campground.

I've been coming to Yosemite since 1951. Susan hiked here in the 1970s. Among my earliest memories are the Firefall, the Merced River, campground black bears, and the elaborate canvas-wall camps erected by our fellow campers. The Firefall was a bonfire lit each summer evening on Glacier Point. As darkness fell, employees of the Glacier Point Hotel (now gone) pushed the blaze over the side so that it fell three thousand feet into the valley, a flaming waterfall. By the late 1960s, the Park Service canceled this spectacle not only because it was inconsistent with preserving the natural state of the valley, but also because the valley vegetation was being trampled by nightly crowds. I recall inner-tubers floating the Merced River. I always wanted to join them but never did because my dad was unwilling to part with his spare tubes. Almost every night a bear or two would wander through the valley campgrounds, raising havoc at some unsuspecting camper's dinner. I envied the elaborate canvas walls California campers arranged around their cots and camp stoves and picnic tables, ad hoc dwellings open to the sky, luxurious it seemed to me, and something we never saw in our rainier Pacific Northwest. But as the 1970s dawned, hard-partying baby boomers began to change the relationship between the Park Service and its visitors. Teaching and explaining—the traditional role of rangers— gave way to law enforcement and crowd control, earlier perhaps in Yosemite than elsewhere. The relationship became tense and adversarial. It's better now, but crowd control remains paramount as officials struggle to protect the valley from the visitors, most of whom are also the taxpayers who fund the park.

A few weeks from now, in May, my San Diego brother will encounter fresh snow and icy roads in nearby Sequoia and Kings Canyon National Parks. That's what to expect in shoulder-season travel. The vagaries of weather means packing for a variety of conditions, which means more stuff. More stuff taxes the Bambi's limited storage space; how you store your stuff determines what inside-your-trailer chaos you'll find when you arrive at your destination.

As in the movie *The Long, Long Trailer*.

In 1953 my parents took my brother and me to see this classic trailering movie. It stars Lucille Ball and Desi Arnaz, then at the height of their fame because of their weekly TV comedy series *I Love Lucy*. The pair play newlyweds Tacy and Nicky Collini, who have opted to buy the eponymous trailer ("long, long" at least by 1950s standards). The trailer is their future home—Nicky is a civil engineer who moves from job to job—and will also be their honeymoon suite. The movie opens in a dark and rainy trailer park as Nicky begins his retrospective tale of trailer travail. What makes the movie so funny and arouses such empathy in RVers ("trailerites," in 1950s lingo) is how each crisis is so familiar: mastering trailer brakes (the installer screaming at a rattled Nicky, "Trailer brake first!"), overenthusiastic trailer-park neighbors (who mob the newlyweds on their wedding night), backing the trailer (in Nicky's case, over Tacy's aunt's prized flowerbeds), an ill-advised attempt to prepare dinner while underway (Lucille Ball at her comedic best amid flying salads, slamming doors, and caroming pots and pans), crossing a Sierra Nevada pass with an underpowered automobile towing an overloaded trailer (due to Tacy's rock collection) followed by a train of angry, horn-honking drivers. For years, my brother and I judged our own family trailer crises by those in *The Long, Long Trailer*. Susan and I now do the same. A few weeks from now, driving up California's Big Sur at fifty miles per hour, a rental-car tourist will stop abruptly in front of us. I'll hit the brakes ("Trailer brake first!"). The Bambi will screech to a halt. When we reach our campground Susan will discover cabinets snapped open and their contents strewn about the trailer like Lucille Ball's runaway dinner.

We have one day left on this Yosemite visit. We decide to drive up to Glacier Point early in the morning. The sun illumes the valley, whitens the glistening granite, shadows the sheer walls, greens the valley floor. El Capitan, Half Dome, Sentinel Dome loom over us. Sierra summits shimmer white with spring snow. We hike up Sentinel Dome for a picnic lunch, sit on a granite escarpment, uncork a bottle of wine, and survey the valley below. John Muir captures how we feel: "The mountains are calling and I must go."

Tomorrow we'll leave Yosemite to head back to the coast. But today we linger atop Sentinel Dome like mountain gods, presiding over Muir's "range of light." The Yosemite we have loved and remembered is still here, still glorious.

CHAPTER 6

DESERT HIGHWAYS, DESERT BYWAYS

Years of drought and famine come and years of flood and famine come,
and the climate is not changed with dance, libation or prayer.

—JOHN WESLEY POWELL

IT'S FOUR A.M. IN the Mojave Desert and somebody is banging on the Bambi's door. I tap Susan on the shoulder to make certain she's awake, then roll out of the trailer bunk, grab a flashlight, and swing the door open. A young woman stands in the moonlight. She's slender, dark haired, wears a black sweater and a short black skirt, and holds a pair of high-heeled shoes in one hand and a purse in the other. "God bless you, sir," she says. "My car is stuck. Can you help get me free?" Indian Cove Campground in Joshua Tree National Park is surreal even in daylight: hills formed by house-sized heaps of granite boulders, spiky desert bushes, campsites tucked into sandy nooks. At night, it's even more so. Tonight a gibbous moon illumes everything bone-white. The campground is deserted. The woman asking for help is definitely not in desert camping attire. Is she alone? Scenarios flash through my mind involving Charles Manson–style cult murders, armed robbery, prostitution, and car/trailer hijackings.

Indian Cove Campground is not quite in the middle of nowhere: The town of Twentynine Palms is only a few miles north, there's a large US Marine Corps desert training base nearby, and the iconic highway formerly known as US 66, a.k.a. "Route 66," runs the length of the park's northern border.

But it's almost nowhere.

Four major deserts dominate the American Southwest: Great Basin, Sonoran, Chihuahuan, and Mojave. Of these, the Mojave is the hottest, the driest, and the lowest (250 feet below sea level in Death Valley) but is also next to the Great Basin, the second-highest, the greater part lying between two thousand and five thousand feet. Most of the Mojave is in California and Nevada, although small portions extend into Utah and Arizona. The range of Joshua trees corresponds to the range of the desert, thus the Joshua is a Mojave "indicator species." The tree looks like a cross between a yucca plant and a palm tree. Mormon settlers named it for what they saw as its resemblance to the prophet Joshua raising his arms in praise to God. Its trunk consists of a fibrous material like yucca, to which it is related. The oldest specimens may live a thousand years. They are pollinated by yucca moths, whose larvae feed on the tree seeds. But this process is controlled by the tree, which can kill its own individual flowers and the larvae on them if the population consumes too many seeds.[1] Joshuas require a winter freeze to propagate and, like many desert plants, will bloom or not depending on the amount of water they receive. This is my first visit to Joshua Tree National Park, although when I was a boy, my family drove by several times when we crossed the Mojave on Route 66, always at night because it was summer and our cars weren't air-conditioned. I remember stopping at Twentynine Palms for fuel and how pleasant the desert night felt: the air dry, the tang of gasoline in the air.

At Indian Cove Campground, four a.m. this October night, the temperature is in the mid-sixties, comfortable, but that's the only thing that's comfortable. I've reluctantly agreed to help the young woman. ("God bless you, sir," she says again.) I tell Susan to dial 911 if

I'm not back in thirty minutes. I tell the young woman I'll follow by car as she walks to where hers is stuck, this a precaution, I tell myself, to keep her from pulling a pistol from her purse. (Of course, she could do the same when we arrive at her car—or a boyfriend could be lying in wait.) She walks down the campground road, illuminated by my headlights, barefooted, wobbling because of the gravel—or is her wobble due to something else? What is it about deserts that engenders idiosyncratic behavior: hermits, prophets, saints, kooks? When we get to the car it's not just stuck, it's *really* stuck, driven over the top of a rock at the end of the campsite parking space. The car is a 1970s-era Mustang convertible with peeling paint. Are those suitcases and cardboard boxes cramming the back seat? "My fiancé and I had a little ruckus," she explains as I dutifully put my shoulder to the car hood. "Last time he came to get me. This time he didn't."

I feel like I'm in a David Lynch movie. My car's headlights flood the scene. The desert night closes around us. The women guns her engine. Sand spews from the Mustang's rear wheels but it doesn't budge. I decide it's time to bring this cinematic event to a close. If the fiancé came once, he might again.

"There's an emergency phone at the Entrance Ranger Station. I'll drive you there," I suggest. The entrance is six miles from the campground. It's also a near a set of suburban-style houses just outside the park gate. The woman (somewhat reluctantly, it seems to me) agrees. While she collects her purse and locks her car, I call Susan with the plan.

The moon has set. Stars spangle the sky. The woman and I drive toward the gate. I can see the lights of Twentynine Palms in the distance. She asks where I'm from, where we're going, thanks me several more times. I almost offer to drive her into town, think better of it, and stop at the emergency telephone. She swings the car door open, looks at the phone, then asks, "How do I use it?" The air seems to have gone out of her.

"Pick it up. The operator will answer."

She steps out. "God bless you, sir."

"Good luck."

She closes the door, turns toward the phone, stops, faces the park gate. I get the feeling she knows the neighborhood. I put the car in gear, press the accelerator, steer out of the parking lot. I don't see where she goes next.

We began our desert loop yesterday on the Pacific coast near Ventura. We navigated the frenetic freeways of Los Angeles County until we escaped into the stark mountains and barren valleys of the desert. A few miles west of Palm Springs we turned north through Morongo Valley, Palm Wells, Yucca Valley, and Joshua Tree, small towns that are economy-class versions of Palm Springs and Palm Desert. As the highway gained altitude, we passed from the Colorado Desert (technically part of the Sonoran) and entered the Mojave, a landscape populated by Joshua trees. Today, the day of our early-morning encounter with the young woman, we're continuing on to Arizona, a drive of over four hundred miles. Our route will cross Joshua Tree National Park, descend to Interstate 10, bypass Palm Springs, span the Colorado River, cross Phoenix (a mistake, as it turns out, because we'll hit rush-hour traffic), and then turn south toward Tucson. We'll climb from only four hundred feet elevation at Palm Springs to over 2,300 feet in Tucson. For nearly the full distance we'll be in the Sonoran Desert, but not until Tucson will it be lush desert.

Lush and *desert* may seem contradictory until you wander through a cactus forest in Saguaro National Park. The park, east and west of Tucson, is the main reason we're visiting. Susan has never been there. I haven't seen it in fifty years. Saguaro are the signature cacti of the movie Wild West: tall, green, tubular, limbs branching like arms. I love their individuality: no two ever the same. As it turns out, plenty of other cacti inhabit the Sonora: organ pipe, senita, hedgehog, claret cup, barrel, fishhook, buckhorn cholla, teddy-bear cholla, beavertail, porcupine, and pancake prickly-pear. Sonoran cacti are astonishingly profuse. Rain is the reason. Summer monsoons drench the Sonora as rising thermals draw moist Gulf of Mexico air into

Arizona. The remnants of Pacific fall hurricanes deliver more rain. Both seasons engender flash floods, lightning, and windblown dust. We've driven I-10 between Phoenix and Tucson when rain fell so hard that we had to pull off the highway because we couldn't see.

Life may be profuse in the Sonora, but it's still desert life. As Edward Abbey puts it in *The Journey Home*, "You will find the flora here as venomous, hooked, barbed, thorny, prickly, needle-toothed, hairy, stickered, mean, bitter, sharp, wiry, and fierce as the animals."

We arrive at Tucson's Rincon East RV Park after dark. We have a reservation and, to our surprise, despite the late hour, a park employee has been awaiting our arrival. He guides us by electric cart to our site, directs us onto a concrete pad, and bids us goodnight. The night is pleasantly warm. Because it's dark, we have no sense of the place, although we know more or less what to expect. Commercial RV parks tend to be utilitarian, designed to accommodate basic needs in a minimum amount of space—power, water, sewer, cable TV, a picnic table. What we'll discover tomorrow is that Rincon East is large, very large. Many of the sites have "manufactured homes," rolled in here once and destined to never roll anywhere else again. Most appear to be vacant, awaiting snowbird owners, presumably still in the north.

The tenants we do see occupy the elderly end of the retiree spectrum. In the early morning before it gets hot and in the late afternoon after it begins to cool, they come out and hobble behind their walkers or buzz about in their golf carts. Susan and I eventually nickname the place "Purgatory." Beyond Rincon's fences are a few suburban homes and then desert. As it turns out, when I go for a swim in the afternoon, I find that the desert has invaded Rincon as well. Seeing me leaning against the swimming pool edge, a passing RVer calls out. "Watch out for the fire ants. They live in the tiles at the edge of the pool."

Paleo-people may have wandered into the Tucson area as early as twelve thousand years ago. From 600 to 1450, the Hohokam lived here, developing centralized villages and extensive irrigation networks and raising crops of maize and cotton. Cotton is still farmed in this area: the Pima Valley variety is considered among the world's finest. Jesuit missionaries arrived in the early 1700s. With Mexican independence, the town, by this time known as Tucson, became part of the Mexican state of Sonora. Americans occupied it in the Mexican War but Tucson did not become part of the United States until 1854 with the Gadsden Purchase, a transaction meant both to secure a railroad route around the Arizona mountains and to ameliorate some of the unjustness of the war. During the Civil War, it briefly served as the capital of the Confederate Territory of Arizona. Then Union troops arrived and drove out the Confederates. The town's late-nineteenth-century history was Wild-West melodramatic, featuring bank robberies, stagecoach bandits, and a shootout led by Wyatt Earp outside the Tucson railroad station. Today the city is the heart of a million-person metropolitan area, site of the University of Arizona, and home to a diversified mix of government institutions and private businesses. It's also a major destination for tourism. In the early twenty-first century, led by business and political leaders, Tucson has undergone a major downtown renaissance.

Susan and I do a morning hike through the Rincon Mountain District of Saguaro National Park. (The park has eastern and western districts separated by the city.) When we get back to the car, it's stifling hot. We lunch in an outdoor Mexican restaurant in the Congress Street Arts and Entertainment District on *moles*.[2] In the afternoon, we visit the Arizona–Sonora Desert Museum.

My first visit to this museum was in the 1960s. What set it apart then was its focus on a specific region, the Sonoran Desert, and its synthesis of museum and zoo, with live animals in natural habitats instead of stuffed animals in dioramas. Since then, the museum has moved to a larger, much more dramatic setting west of Tucson Mountain Park. There's a large parking area, several mission-style buildings, and a network of paths to various desert habitats: Cat Canyon (mountain lions and ocelots), the Desert Loop Trail (coyotes, javelinas, lizards), the Riparian Corridor (river otters, beavers, coatimundis, native fish), and our favorite, the Hummingbird Aviary, where dozens of hummers whiz back and forth like airborne emeralds, rubies, and sapphires.

The afternoon begins to cool. From the museum veranda, below and behind us in the afternoon sun, saguaro glow like golden candlesticks. To the west, the Pescadero and Recortado Mountains rise. The geology here is basin and range, as it is in Nevada—wide valleys, steep mountains. But the desert climate shapes its own geology. Geologists have developed a unique vocabulary to describe it, much adopted from the language of the original European explorers: aprons of alluvial outwash material (sand, clay, gravel, boulders) called *bajadas*; *arroyos* that flood periodically and sweep away cattle and even automobiles during torrential rains that vanish into the dry soil as soon as the clouds move away; *dust devils* that spiral skyward on all but the coolest days; fierce *haboob* dust storms that turn day into night; *desert varnish*, a shiny black or orange-yellow coating that forms when clay, iron, manganese, and rainwater combine; *spheroidal weathering*, the process when granite is subjected to extremes of daytime heat and nighttime cooling and converts from flat-faced blocks to rounded boulders; *caliche*, hardpan that forms as calcium carbonate and other soluble minerals collect and then inhibits water permeability and root growth.

It's a hard geology for a hard land.

After Tucson, we head north. We leave the Sonoran Desert, cross Arizona's Central Highlands, pause at the rim of the Colorado Plateau. Once again, we must drive the Bambi through Phoenix. I like the Sonoran Desert and I like Tucson but I have less affection for Phoenix. Four million people live in this improbable desert metropolis, whose continued survival depends on air conditioning and water. How long the electricity and the water will last, nobody knows. As Edward Abbey writes in *Desert Solitaire*,

Water, water, water. . . . There is no shortage of water in the desert but exactly the right amount, a perfect ratio of water to rock, of water to sand, insuring that wide, free, open, generous spacing among plants and animals, homes and towns and cities, which makes the arid West so different from any other part of the nation. There is no lack of water here unless you try to establish a city where no city should be.

By lunchtime, we're north of the city and having a picnic at the Sunset Point Rest Stop in—such a lovely name!—Bumble Bee, Arizona.

You can simplify Arizona's geology into three provinces: basin and range in the south; a Central Highlands across the middle; the Colorado Plateau to the north. The highlands meet the plateau at an escarpment called the Mogollon Rim (pronounced MUG-ee-yun). Here, red-rock formations front the desert and the land abruptly rises. The rim is where we're bound. My San Diego brother has agreed to meet us in Oak Creek Canyon in the Coconino National Forest, just north of the spa, golfing, and New Age community of Sedona.

Where Sedona, at four thousand feet, lies at the bottom of the rim, the town of Flagstaff, at over seven thousand feet, perches at the top. If you follow I-17, bypassing Sedona, you may only be generally aware of how much elevation you're gaining, but you'll certainly notice how the desert flora gives way to pine forest. But if you follow Arizona Alt 89 through Sedona and up into Oak Creek Canyon, the ascent is perfectly apparent. The road follows Oak Creek, then climbs the rim wall in a series of heart-stopping switchbacks. Our campground, Cave Spring, is just south of where the switchbacks begin.

Cave Spring this October afternoon blazes with arboreal color— reds, browns, yellows, golds. A concessionaire runs the campground for the National Forest Service. We check in. Oak Creek Canyon is a popular Arizona destination, some say second only to the Grand Canyon. Today the campground is only three-quarters full. My brother and sister-in-law have already arrived. We tour their new Lance trailer, far more commodious than our Bambi. It includes a stand-up refrigerator, an oven, electric jacks. The four of us trade RV stories, build a campfire, and dine outside. The last time my brother and I trailer-camped together, we were kids.

The next day we hike the Devil's Bridge Trail and ascend a natural arch in red sandstone. We lunch in Sedona, have dinner again at our campsite. What we see in the white limestone of the Oak Creek Canyon and in the red sandstone formations around Sedona foreshadows what we'll see farther north.[3] My brother and his wife will head south to Tucson tomorrow. Susan and I will ascend the Mogollon Rim and enter a new geological domain—the Colorado Plateau, a region so different, so beautiful, so remarkable that it deserves its own chapter.

CHAPTER 7

THE COLORADO PLATEAU

US 89, 89A, and the Carmel Byway

You have to get over the color green; you have to quit associating beauty with gardens and lawns; you have to get used to an inhuman scale; you have to understand geological time.

—WALLACE STEGNER

SUSAN AND I HAVE just unhitched our Bambi trailer in the Navajo Nation's Monument Valley Visitor Center parking lot. We're on the Colorado Plateau and we're traveling into deep geological time. The temperature is in the low eighties, the soil rust-red, the sky cerulean blue, the sun, at three p.m. this October afternoon, low on the horizon. Ten minutes ago, we purchased a self-drive pass for the seventeen-mile loop drive: no trailers permitted. A wide sagebrush plain spreads before us. Orange sandstone buttes rise from the plain like islands in the sky.

We feel like characters in a John Ford western. Ford filmed many movies here: *The Searchers, How the West was Won, She Wore a Yellow Ribbon, Stagecoach, Rio Grande,* and others. A lot of not-John-Ford movies have also been filmed in the valley: *Forrest Gump, 2001: A Space Odyssey, Easy Rider, Butch Cassidy and the Sundance Kid, Back to the*

Future III, *Mission Impossible II*, *National Lampoon's Vacation*, and the HBO series *Westworld*. Monument Valley has become a cinematic icon for the American West.

The guided tour might have been a better choice. I'm so busy trying to avoid bus-size potholes and muddy washouts that I don't have time to appreciate the monoliths that tower over us. When I nearly high-center the MDX exiting a dry wash, we decide to turn around.

Monument Valley straddles the Utah–Arizona border, not far from Four Corners, where Utah, Arizona, New Mexico, and Colorado meet. It's also the approximate center of the Colorado Plateau, whose unusual geological history is why so many glorious National Parks and Monuments have been created here, among them Grand Canyon, Arches, Bryce, Zion, Canyonlands, Capitol Reef, Mesa Verde, Canyon de Chelly, Cedar Breaks, Natural Bridges, and Grand Staircase–Escalante. This geology is so extraordinary that an entire lexicon has been created to describe it: *mesas, domes, reefs, hoodoos, goblins, fins, natural bridges, slot canyons, alcoves, windows, hogbacks, arches.*

And here's a paradox. It's not a geology of cataclysm but of stability.

As tectonic plates crashed elsewhere in their worldwide four-billion-year bumper-car derby, here on the Colorado Plateau geology unfolded more sedately. Mountains rose and weathered to flat plains. Rivers braided swampy coastlines. Seas flooded in and ebbed out. Deserts formed and became seas again. Mountains rose one more time. Each of these phases lasted tens of millions of years over a term of nearly two billion years, perhaps because the earth's crust was thicker under the Colorado Plateau and moderated the forces of tectonic chaos. Whatever the reason, the materials deposed were left largely intact, a layer cake of geological history. Was it dry or wet or windy? Shallow seas or coastal deserts? Protozoa or dinosaurs? The layers tell the tale. What's more, the age and material of each layer led to a wide variety of geological forms, each eroding in a different manner: steep cliffs (sandstone and limestone), crumbly skirts and talus benches (mudstone and shale), resistant caps (igneous basalt and granite), or combinations (hoodoos, arches, natural bridges, reefs).

In Monument Valley, three layers make up the towers or "monuments": The bottom is Organ Rock Shale, the middle, and tallest, De Chelly Sandstone, the top the Moenkopi Formation (with a cap of Shinarump Conglomerate). In the De Chelly Sandstone you can see what geologists call *eolian* cross-bedding—thin, oblique, wind-generated layers formed when this part of the plateau was a dune desert. Every layer has its own story, its own prime time in the planet's history.

Susan and I hook up the trailer at the visitor center.[1] We spend the night at a nearby RV park that belongs to the Navajo Nation. Our site lies below red-rock sandstone cliffs. I take a picture of Susan walking to the showers followed by a parade of black house cats. The evening is warm and still. Night falls. Coyotes howl at a crescent moon. After midnight a chill wind begins to blow. It's almost November and we're at five thousand feet; we must expect weather less predictable.

The next day we head to Utah's Zion National Park. We backtrack south on US 191 into Arizona, then head southwest at Kayenta on US 160. We turn north on Arizona 68 (bypassing Navajo National Monument and its cliff dwellings, which we make a note to visit on a future trip), then pass through Kaibeto, intersect with US 89, and arrive in Page with just enough time to shop and have lunch. We cross the Colorado River below Glen Canyon Dam, a controversial hydropower/ water project that filled Lake Powell, drowned Glen Canyon, and changed the Colorado's flow, silt load, and temperature and that may someday have to be removed due to silt buildup behind the dam.[2]

Once we cross the river, we enter the Grand Staircase–Escalante National Monument. This owes its name to its series of colorful cliffs—major escarpments that run parallel for dozens of miles on an east–west line, gaining altitude south to north, each set of cliffs eponymous for their color: Chocolate, Vermilion, White, Gray, Pink. South and west from us, we can see the Vermilion Cliffs, the second set in the staircase. These belong to the Moenave Formation, dating to the Jurassic Period, once a river-braided coastal plain where dinosaurs, other reptiles, amphibians, and early mammals lived. Iron oxides and manganese create the intense red and blue colors. This layer and those of the other Grand Staircase–Escalante cliffs extend under the plateau to the north: Because the plateau rises to the north, we'll encounter the Moenave again at the bottom of Zion Canyon. From here to Zion, all the layers were deposited during the Jurassic and Triassic Periods (the early to middle Mesozoic Era), 145 million to 252 million years ago.[3]

US 89 skirts the base of the Vermilion Cliffs. At Kanab, a town with neat houses, shade trees, and gridded streets—what you can expect in Utah's Mormon-settled towns—we turn north on the Zion–Mount Carmel Byway, ascending to the base of the White Cliffs (Temple Cap and Navajo Sandstone, both Jurassic). The White Cliffs mark the third escarpment in the Grand Staircase. At Carmel Junction, we head west on Utah 9. Strobe-lit signs signal deer crossings. We see plenty of carcasses before we reach the Zion Park east entrance.

We must purchase our ticket before four o'clock because the Zion–Mount Carmel Tunnel—a wonder of 1930s highway engineering and the only eastern route into Zion Canyon—is open only intermittently for oversize vehicles like our Bambi. When we reach the tunnel, a park ranger signals us through. The tunnel has windows on the canyon, but these days, with more traffic, stopping is no longer permitted. After we emerge, we descend switchbacks into Zion Canyon.

Where the Grand Canyon overwhelms you, Zion encloses you. The canyon is compact, never more than three thousand feet in width, deep, its rim rising a half mile above us, and colorful, its walls layered with red, brown, pink, salmon, and white stone. Once we reach the canyon floor, it's pastoral—the Virgin River, which originally cut the canyon, snakes down its course with, on this October afternoon, its riverbank cottonwoods blazing gold.

A Mormon pioneer, Isaac Behunin, named this place. In Mormon theology, Zion is where members of the faith will gather to await the Last Judgment. My grandmother, Catherine Seybolt, was not Mormon, but she was the widow of a Methodist minister and she loved Zion National Park. In 1957, when she was seventy-five years old, she and our family camped here. She loved it not only for its biblical name, but because she thought that heaven—"Paradise" is what she liked to call it—might look a lot like Zion.

The vehicle-accessible part of the canyon is less than twenty miles long. At Canyon Junction, site of the road to the Zion Canyon Lodge and beyond, unless you are a lodge guest, you may only hike, bike, or take a shuttle bus.

Tonight, the electrified Watchmen Campground is full, so we take a site in the South Campground. Susan and I ride our bikes up the Par'rus Trail, which begins at the South Campground and runs up-canyon to Canyon Junction. The paved trail crosses back and forth over the Virgin River. Cottonwoods, tamarisk, willows, maple, cacti, and sagebrush line the riverbank. We pass a young female ranger leading a tour of schoolchildren. Showers threaten. We opt to cycle back to our campsite and board the shuttle bus to Zion Lodge. The bus isn't crowded but on a subsequent trip, hundreds of people will be waiting at the visitor center stop.

Zion has become a much-visited park. If a "grand tour" of Western National Parks exists, Zion would be on it, not only for its own beauty but for its proximity to other parks, the famous Southwestern parks like Grand Canyon and Bryce Canyon and also Yellowstone and the Grand Tetons. Zion is just off Interstate 15, a major auto route from Los Angeles and Las Vegas to Yellowstone.

Our last day dawns clear and bright but with an autumnal chill. Clouds loom to the south. We decide to hike. The trail follows a dry creek. I snap a picture of Susan. The cliffs behind her tell the geological story of Zion far back into deep time. Zion Canyon consists of nine sedimentary layers. Navajo Sandstone is the thickest and dominates its walls. Its range of color—red, mauve, pink, and white—originates from the varying trace minerals in the stone, mostly oxides of iron. Where did so much sandstone come from? The answer lies in a wonderful geological detective story.

In the late Jurassic, after the supercontinent of Pangaea broke apart, ancestral North America began to assume its present shape. On its western shore, a desert had formed, larger than today's Sahara. Wind-generated dunes rose to several hundred feet. Erosion on the dunes' windward sides and rainfall on their leeward sides formed ripples in the sand. Over millions of years the sand amassed to over two thousand feet, most lying below younger layers. Mineral-laden groundwater, calcium carbonate in particular, circulated through the sand and fused the grains together. This created sandstone, but the dune ripples remained, frozen in crossbedded stone, in effect frozen in time, their pattern revealing the prevailing wind direction that blew when the dunes formed. The sand grains—fine, well-sorted, and well-rounded—indicated a long, wind-blown journey.

On the eastern littoral of North America at the time, the Appalachian Mountains had risen to elevations higher than the modern-day Himalayas. Rivers flowed down their western slopes carrying eroded sand, including a rare and durable mineral, zircon, which also contains elements of uranium. Because of its uranium and because of its durability, zircon can be dated to the original time of its crystallization. As the zircon and other sands deposited on the river banks, the wind picked them up and swept them away, tumbling and bumping, until they dropped on the desert that would in time become Navajo Sandstone. The dates of its zircon crystals match those of the Appalachian bedrock. What I love about this story is how the small details—the crossbedded wind pattern in stone, the tumbling grains of zircon, the uranium dating—illuminate a continental big picture.

Each layer of stone tells a unique history: so many layers, so many stories, so much time. Maybe time is the plateau's primary geological message. Most of us don't come here for history or geology lessons. We come here for beauty, for a land different from home: drier if we come from a wet place, warmer if we come from a cold place, cooler if we come from a hot place. We come here because we like the smell of sagebrush and piñon pine, the spring wildflowers, the autumnal gold of cottonwoods, the stubborn persistence of cacti. We come here to climb sandstone cliffs, to hike switchback trails, to cycle the canyon floors. If we learn anything, perhaps what we learn is accidental, a gift.

Susan and I depart Zion and head north up Interstate 15. We pass through Salt Lake City, spend a chilly night at the Golden Spike RV Park, a few miles from where the Central Pacific was joined to the Union Pacific, completing America's first transcontinental railroad, and where our water hose will freeze and where, in the morning, I'll forget to retract one of our stabilizing jacks so that the jack will crumple as I attempt to tow Bambi away and, in the course of trying to recover, I will poke a hole in the bottom of the trailer.

So goes the trailering life!

But we'll return to the Colorado Plateau. Time and again, we'll be back for another voyage into time.

CHAPTER 8

VOLCANIC LEGACY

US 97, Maryhill to Mount Shasta

Glorious, stirring sight! The poetry of motion! The *real* way to travel!
The *only* way to travel! Here today—in next week tomorrow! Villages
skipped, towns and cities jumped—always somebody else's horizon!

—KENNETH GRAHAME

WE'RE STANDING IN STONEHENGE above the Columbia River
in Washington State. What we're doing here you can't do at the real
English Stonehenge: that is, go inside. Here at Maryhill, anybody can
wander inside. Sam Hill, railroad man, pacifist, and son-in-law to
James Hill, founder of the Great Northern Railroad, built this replica
and also the nearby mansion that became the Maryhill Art Museum.
Hill originally bought the land in order to lure Quakers to a pacifist
colony. When no Quakers came, he decided to build a memorial to
the soldiers then dying in World War I. He chose to replicate Stonehenge because at the time people believed the legend (now discred-
ited) that the original site had been a place of human sacrifice; Hill
intended his monument to remind people, as he wrote, that "human-
ity is still being sacrificed to the god of war." The Maryhill Stonehenge
is smaller than the English Stonehenge but at the summer and winter

solstices, light falls exactly as it does at the real Stonehenge. Signs warn visitors to smoke only in their cars due to sagebrush fire hazard and to watch for rattlesnakes. Now, in early June, the sagebrush and grasses are spring green. While there is little fire hazard, you still have to worry about the snakes.

Susan and I haven't set a route for this Bambi trip, not even a date to return to Seattle. This time we're playing it loose. We'll take a relaxed drive down Central Oregon via US 97 amid spring wildflowers and greening wheat fields, with a visit to Newberry National Volcanic Monument (a place we've passed but never stayed), perhaps a stop at Crater Lake, too (if the weather's right and the campground is snow-free), a day or two lounging along the Rogue River, maybe even a return up the coast.

This journey will be slow going, with no set schedule: the ideal of serendipitous trailering.

"Trailering" (or "caravanning," as the British call it) goes back a long way, to the Romani, sheepherders, snake-oil salesmen, and itinerant evangelists. The world's first leisure trailer, aaccording to Wikipedia, was built by the Bristol Wagon and Carriage Works in 1880 for Dr. William Gordon Staples, who ordered a "gentleman's caravan." Staples named it *Wanderer* and traveled around the British countryside, later writing a book about his travels called *The Gentleman Gypsy*. Thirty years later, in Kenneth Grahame's *The Wind in the Willows*, Toad's first obsession is a horse-drawn caravan and he exclaims to Rat and Mole about the "poetry of motion." In the United States, a thriving travel-trailer industry was wiped out by the Great Depression, only to resurrect itself in the decade after the Second World War. An alphabet of manufacturers flooded the market: Airstream, Aljo, Chinook, Coachman, Fleetwood, Nomad, Oasis, Scotsman, Shasta, Terry, Winnebago, names synonymous with a nation hitting the highways. By the early 1960s, as John Steinbeck described in *Travels with Charley: In Search of America*, when his neighbors scrutinized his camper, "I saw in their eyes something I was to see over and over in every part of the nation—a burning desire to go, to move, to get under way, anyplace, away from any here . . . nearly every American hungers to move."

What we don't know yet is whether two phenomena, the fast-retiring baby boom generation and the 2020 COVID pandemic, will soon vanquish serendipitous trailering. Will RVers flood the campgrounds, setting off a year-round scramble for reserved sites, thus ending the early spring and late fall "off-season" when reservations have rarely been necessary?[1]

Susan and I lunch at the Maryhill Winery. The winery sits on a rock shelf a thousand feet above the Columbia. We can see up and down river. Pale green bluffs to the east give way to black cliffs to the west, fluted by basalt columns. The river flows slowly here, tamed by a series of dams. Salish peoples once netted salmon at Celilo Falls, just downriver, as recently as your parents' or certainly your grandparents' lifetimes. The falls are underwater now.

The Columbia River Gorge lies immediately west. The river flows through a water gap in the Cascade Mountains. For the next eighty miles, basalt cliffs rise to four thousand feet, waterfalls plunge from hanging valleys, and the climate changes within a few miles from dry grasslands to temperate rainforest. The gorge creation is a story of geologic catastrophe: an ocean of erupted lava, the world's greatest floods. It's also a story of the steadfastness of the Columbia.

Here's how it goes:

About twenty million years ago, as the Pacific Ocean Plate dove under what is now Washington and Oregon, the ancestral Cascade Mountains began to rise. Geologists don't really know how old the ancestral Columbia was. But sixteen-and-a-half million years ago, when a series of volcanic eruptions began in the area where Washington, Oregon, and Idaho meet, the Columbia River began evolving, shaped by those eruptions. The lava emerged from northwest-trending vents, often many miles long; it was unusually fluid; and the eruptions continued off and on for another seven million years. In times of maximum activity, it changed the global climate and covered eastern Washington, eastern Oregon, and western Idaho with layered basalt several thousand feet deep, basalt so heavy that the entire Columbia Basin tilted down to the southwest. Ultimately, the lava flowed all the way to the Pacific. The probable cause of this volcanism was a plume of magma rising through the Earth's crust—a "hot spot"—one that geologists believe now centers under the Yellowstone Volcano in northwest Wyoming. The plume didn't move. The North American Plate moved to the west, over the plume. It's also possible that the eruptions were triggered in conjunction with the East Pacific Rise, a spreading center (like the one under the Atlantic) now under the continent. As the flood-basalt volcanism waned, the rivers were able to establish courses. When the modern Cascades began to erupt five to seven million years ago, the river found a way to continue its journey to the Pacific.

Then, twelve thousand years ago, at the end of the last Ice Age, another series of cataclysms swept the Columbia. The Cordillera Ice Cap then covered most of western Canada and much of the northwestern United States. Ice Age lakes had formed in what is now Montana. The lakes were equivalent in size to today's Lake Superior but, unlike Superior, only ice dams held them in place. Recall that ice, no matter its mass, floats like the cubes in a gin and tonic when the water around it becomes deep enough. As the climate warmed and the ice cap melted, the lakes grew deeper, the ice dams lifted, and the dams suddenly and repeatedly collapsed. The largest floods the planet has ever seen swept across Montana, Idaho, Washington, and Oregon, gouging the Columbia Basin with torrents of water carrying with it topsoil, gravel, forest debris, and icebergs, floating boulders as big as houses. This bulldozed a network of "coulee" canyons throughout the Basin. It excavated the Columbia's present river course and created today's gorge. This didn't happen once but several dozen times as the ice dams re-formed, the glacial lakes refilled, and the dams failed again. What you see today in the Columbia River Gorge is the legacy of cataclysms: Miocene volcanic fire and Pleistocene Ice Age ice.

We can see Mount Hood rising beyond the gorge, a white wizard's hat, one in the chain of great Cascade stratovolcanoes, born in fire, shaped by ice, like the gorge itself, although a different kind of fire, a different kind of ice. Tomorrow, when we head south on US 97, we'll pass under six of these ice-and-snow-covered, fire-birthed giants. The Columbia Basin flood basalts, as large as they were, erupted at leisurely intervals, perhaps twenty to twenty-five thousand years between events. But only three decades have passed since Mount St. Helens blew its top, only a century since Mount Lassen erupted, only 7,700 years since Mount Mazama exploded and left behind Crater Lake and laid hundreds of thousands of tons of ash across the Northwest.

We get underway early from Maryhill. We cross the Columbia River, enter Oregon, and ascend US 97 out of the gorge and up onto the Umatilla Plateau. The highway climbs through a narrow, basalt-walled canyon until it reaches the rolling fields of winter wheat, alfalfa, and barley atop the plateau. Here, atop basalt, the loess soil laid down by Ice Age winds thins and the grain fields give way to rangeland. We pass through Wasco, DeMoss Springs, and Moro. The sun shines. Wildflowers brighten the highway margins; the world feels Edenic. In the distance, the first of the Oregon volcanoes appear: Hood and Jefferson, white-clad patriarchs above the Cascade crest. US 97 briefly breasts the Deschutes River Canyon, where the dry, treeless highlands are covered by wild grasses and sagebrush, but along the river's bank alder, mock orange, clematis, and chokecherry shadow the flow. We're in the high-desert ecosystem now and also the province of volcanism. To the east, the hills give way to palisades and buttes. To the west are more volcanoes—Mount Washington, the Sisters, Bachelor. We're gaining altitude. Ponderosa pine replaces hay and sagebrush. At the town of Redmond, car dealers, shopping malls, and fast-food franchises replace pines. Redmond and Bend have grown fast in the twenty-first century and are now home to more than two hundred thousand inhabitants. For years, this was the commercial heart of Central Oregon's logging and ranching industries. Now Redmond and Bend have become a hub for fly-fishing, skiing, cycling, river rafting, and retirement. The climate is dry, not too hot in the summer, not too cold in the winter. The Deschutes flows through the towns, bordered in Bend by city parks, bikeways, and handsome late Victorian homes. Brewpubs and upscale restaurants have opened. As we enter Bend's city center, traffic comes to a dead stop. Several people are escorting a fawn off the freeway.

We don't stop in Bend this trip. Our goal today is Newberry Volcano.

Newberry began erupting fast-flowing lava about four hundred thousand years ago, ultimately covering an area the size of Rhode Island and creating the largest of the Cascade volcanoes by volume.[2] More recently, Newberry emitted explosive, high-viscosity lavas. The last of these eruptions occurred less than 1,500 years ago. The other defining characteristic of Newberry is its caldera, formed when the magma chamber below its summit collapsed. The caldera now harbors two lakes. Geologists have identified a magma chamber below Newberry with temperatures high enough to provide geothermal energy. You can drive through Bend and down US 97 and never realize that the low ridge to the east is a volcano. Because two hundred thousand people live in the direct path of historic eruptions, Newberry is routinely monitored by the National Geodetic Survey.

We stop at the Lava Lands Visitor Center and discover that car/trailer rigs over twenty-two feet are prohibited in the regular parking area. No problem. The gate-keeper dispatches us to the RV parking area, a short walk from the center exhibits. We have a leisurely lunch followed by a quick tour of the visitor center. We learn that Newberry Crater has over four hundred vents and cones, the highest of any single Cascades volcano. We also learn that unlike the others in this range, plate subduction is not the direct cause of Newberry's volcanism. Nor is a hot spot. Transverse movement along the east–west Brothers Fault Zone generates the heat that melts the rock that creates the magma that lies under Newberry.

We continue south. At La Pine, we exit US 97 and commence the climb to the volcano's caldera. Newberry's slope is so gentle that we don't feel as though we're ascending a volcano, even though we began at four thousand feet and the place we'll camp lies at six thousand. Ponderosa gives way to hemlock, firs, and white-bark pine. Newberry's ecosystem transitions to High Cascades from high desert. Its summit, Paulina Peak, rises to nearly eight thousand feet. We enter the caldera. Snow whitens the black-lava rim. A gray cinder cone separates twin lakes, East and Paulina. We check out the campgrounds, two on each lake with an RV park with electricity set back from both. We score a spectacular spot in Little Crater Campground, on the Paulina Lake shore, adjacent to a boat-launching ramp. The crater rim wraps the lake. Pines and spruce shade our Bambi. It's a perfect site and we're lucky to get it. In our many years of sailing, Susan and I always designate the best anchorage of a voyage "the one perfect harbor." Surely Paulina Lake will be the "perfect harbor" of this Bambi voyage.

We linger for a couple of days, loaf in our campsite, hike around the lake, drive down to Sunriver Resort, dine in a brewpub. The weather begins to turn. Clouds roll in and a chilly wind begins to blow. We decide it's time to hit the road. We need cell phone coverage to determine what the weather will be at Crater Lake.

Mount Mazama, a twelve-thousand-foot volcano, once stood in place of Crater Lake. Mazama began rising about four hundred thousand years ago. Eight thousand years ago, it blew its top in an eruption forty times larger than the 1980 Mount St. Helens event. A mile-wide column of volcanic debris shot five to ten miles skyward at twice the speed of sound. Flow after flow erupted. The top of Mazama began to sink into the emptying magma chamber. When the cycle finally ended, a caldera had formed that would fill with rain and snowmelt and became Crater Lake. A half-million square miles of the Northwest, including Oregon, Washington, California, Idaho, Montana, Utah, Nevada, Wyoming, British Columbia, Alberta, and Saskatchewan, were blanketed by Mazama tephra (ash and debris). Geologists came to call this "Mazama Ash" and it has become a useful dating tool. If you know where something is relative to Mazama Ash, you can establish its date relative to when the volcano erupted. In Newberry Crater, archeologists were able to date the presence of people because the foundations of Native American dwellings were buried in Mazama Ash.

As Susan and I descend from Newberry Crater, we also descend into cell phone range. We check the Crater Lake National Park weather forecasts. What we learn is that the weather will be bad—snow mixed with rain. One campground is open but not snow-free, and the highway on which we planned to enter the park is under construction. In the pre–cell phone era, we would have winged it, maybe enjoyed the unexpected beauty of Crater Lake in the snow or the pleasure of an empty campground. But we blow it off.

Cell phones kill serendipity, too.

Can we book tickets to a Shakespeare play in Ashland instead? We can. We do.

For now, we bid goodbye to US 97, a highway that defines our volcanic legacy. If, like Susan and me, you grew up in the Northwest, stratovolcanoes overshadowed your youth. We camped in their meadows, skied their slopes, hiked their trails. They defined our Northwest identity. The Cascade volcanoes are sleeping, not dead. They tower over our cities. They are dangerous. They are old friends. We know them in all seasons. They will erupt again.

CHAPTER 9

YELLOWSTONE

The headwaters of the Yellowstone River . . . is hereby reserved and withdrawn from settlement, occupancy, or sale . . . and dedicated and set apart as a public park or pleasuring-ground for the benefit and enjoyment of the people.

—THE YELLOWSTONE NATIONAL PARK
PROTECTION ACT OF CONGRESS (1872)

YELLOWSTONE WAS THE FIRST US National Park and can reasonably lay claim to being the first national park in the world. It is the second-largest park in the lower forty-eight states (only Death Valley is larger) and is the fifth-most-visited (behind Great Smoky Mountains, Grand Canyon, Rocky Mountain, and Yosemite). It's famous for its hot springs and geysers, for its Grand Canyon of the Yellowstone, for Yellowstone Lake, and for its diverse wildlife—especially bison and bears. Most of the park is in Wyoming, although slices lie in Idaho and Montana. The Continental Divide runs through the park.

I first visited Yellowstone in 1950. I was three years old. Now, sixty-some years later, what I remember from that visit is likely based on my dad's Kodak slides rather than any actual memories. But in August 1956, we visited again, spending our last day here at Mammoth Hot Springs. It snowed. So much steam rose from the hot springs that it was difficult to see the terraces. I remember the sulfur pong and I remember buying "real" cowboy hats at the Mammoth Hot Springs store with my brother Charlie and I remember the feel of the hat felt in my hands and the smell of the leather band. Charlie's was black, mine white, how we saw ourselves in the family dynamic. Years later, in 1997, on a chilly June afternoon, Susan, our son John, and I tent-camped in Yellowstone's Canyon Campground. It snowed on that visit, too.

We have entered the northeastern park entrance at Mammoth Hot Springs. The hot springs terraces, campground, lodge, stores, visitor center, and staff buildings are located a few miles inside the park entrance. We won't camp here today. Our goal is the more centrally located Canyon Campground near the Grand Canyon of the Yellowstone. We do plan to stop for lunch and take a short hike through the terraces—if we can find parking. The lots below the terraces are not trailer-friendly. But adjacent to the upper terrace, outside a wooden corral fence, we find a large, unmarked gravel parking area.

Sulfur pongs the air. Wooden boardwalks mark the trail. Water gurgles down the terrace steps. Most Yellowstone hot springs are clear deep pools rising through volcanic rhyolite lava. What makes Mammoth's hot springs different is limestone laid down in an ancient sea. This results in a staircase of white travertine, calcium carbonate, the same chemical in antacid tablets. Algae and minerals dye the travertine yellow, orange, and brown. Where the flow slows, flat staircase terraces form, almost like the steps in a Mayan temple. In a faster flow, lacy faces calcify, like stone icicles. The trails aren't crowded but we notice something we never have during previous visits. We are hearing a Tower of Babel: Our fellow hikers speak in German, Swedish, Dutch, Russian, Italian, Spanish, Japanese, Chinese, Polish, French. America's National Parks have become the world's national parks.

Our goal this September day is Canyon Village. We'll follow a clockwise route outside the geothermal part of Yellowstone, through high open grassland where elk, bison, antelope, bear, and deer are common. We spot a solitary bison and pull off the road to snap a photo. Bison are big, fast, and dangerous. For us, a car snapshot will suffice. We ascend Dunraven Pass to nearly nine thousand feet. After the pass, we cross into the Yellowstone Caldera. Unlike Paulina Lake at Newberry Crater or Crater Lake, we don't feel we're entering a caldera. The scale here is too large, the rim too eroded over time. But much of the park is sited within the caldera. The first Yellowstone caldera occurred 2.1 million years ago, blowing six hundred cubic miles of material into the atmosphere. (By comparison, Mount St. Helens blew only a tenth of a cubic mile of material.) The third caldera was created six hundred and forty thousand years ago, after the volcano erupted 240 cubic miles of material. Both events would have changed the earth's climate. Both would likely have caused many plants and animals to become extinct. It's no surprise that the Yellowstone Volcano is one of the most monitored volcanoes in the USA.

When I visited Yellowstone as a boy, I don't recall the terms *volcano* or *caldera* even mentioned. I don't recall it being described as a volcano at all. We were told, of course, that hot springs and geysers were the result of heat generated beneath the earth's surface, but how that heat was generated, at least as I remember the exhibits, was not further explained. In the 1960s, however, geology was undergoing a revolution. Plate tectonics was rewriting the story. Volcanism could now be explained as the result of melting magma generated by ocean plates spreading and then subducting under continental plates or an oceanic or continental plate that sailed over a "hot spot," a fixed magma plume that for some reason or other rose up from a particular spot to the planet's surface, leaving in the plate's trail a series of volcanic events, like the chain of Hawaiian islands or, in Yellowstone's case (perhaps), the Columbia Basin basalts, the Snake River floodplain basalts, and Yellowstone itself.[1] The 1980s eruption of Mount St. Helens changed the popular perception of what an active volcano was, given the vividness of the peak blowing itself apart on TV.

Dormant no longer meant dead. In fact, it may never have meant anything at all.

We arrive at Canyon Village. Fuel. Food. Showers. Laundries. Restaurants. Cabins. Campgrounds. A functional, car-centric architecture the Park Service ambivalently calls *Park Service Moderne*. We've made a reservation at the Canyon Campground, so we bypass the lodges, stores, and supermarket. The campground registration turns out to be in a large building labeled "Camper Services" that also houses a laundromat and pay showers. It looks like a school gymnasium. In the past, Park Service employees ran the campgrounds and concessionaires ran the lodges. Now, especially in the larger parks, concessionaires have taken over campgrounds as well.

A grandmotherly baby boomer checks us in. "Spirits are high," she explains. "We're partying before we head out to where we spend our winters."

She briefs us on the regulations. Generators are permissible from eight a.m. to eight p.m. (We won't run ours at all.) No food or cooking gear may be left outside or within sight inside vehicles; even tablecloths must be removed from picnic tables between meals. This is about bears. Pets must be leashed (also about bears). Inspections will be conducted and fines levied if offenses occur (again, about bears). Over time, the Park Service has grown more concerned about its resident carnivores. When I visited Yellowstone as a kid, we always saw bears along the highway and in the campgrounds. We even drove to the campground garbage dumps to watch the bears feed in the evening, a then-Yellowstone tradition.

Even with bears less visible, they discomfit us, especially when we're hiking. Susan and I don't plan to do any overnight backpacking, but we want to take a day hike or two. The Park Service offers this advice. Hike in groups. (Are two people a group? The Park Service says "no.") Make noise. (Bear bells are no longer considered effective, they say; try singing.) Carry bear spray, a pepper compound intended to blind an attacking bear. (How close to the bear do you need to be in order to blind it?) Back up slowly if you encounter a bear. Don't even bother to run or climb a tree; bears can outrun people and climb better than you do. If a bear continues to attack, fall to the ground, assume a fetal position, and maneuver your backpack between you and the bear.

We find little of this encouraging.

The check-in grandmother assigns us site number 215. Canyon Campground consists of multiple loops in a dense lodgepole pine forest. There's no undergrowth. At this altitude, the pines are skinny and no taller than twenty or thirty feet. Site 215 is a back-in, spacious, with a picnic table and a fire ring and plenty of space between us and our nearest neighbors.

Tomorrow, we plan to hike along the rim of the Yellowstone Grand Canyon. We decide to spring for a can of anti-bear pepper spray—fifty dollars!

The Grand Canyon really is yellow, how Yellowstone got its name. The yellow comes from sulfur, but there are other colors too—whites, grays, beiges. Two glorious waterfalls lie on the river's course, one of which, the three-hundred-foot Lower Falls, is as symbolic of the nation's parks as Half Dome in Yosemite or Old Faithful Geyser here in Yellowstone.

Tourists crowd the Canyon Rim Trail to Artist Point, including the steep steel stairs down to the Lower Falls viewpoint. A group of boisterous Russians push past us, the men physically fit with tight T-shirts and short-clipped haircuts, the women bored, blonde, and blasé in their disco makeup. The stairs are called Uncle Tom's Trail, Uncle Tom Richardson being an early park concessionaire who guided visitors into the canyon by a network of rope ladders. A fellow hiker offers to snap our picture at the falls viewing platform. Later, when I email the photo to my brother in Seattle, he replies that he was at the same place just a couple of weeks earlier.

Beyond Artist Point the crowds thin out. *Really* thin out. Is this about bears? I check my bear-spray can—yes, still hanging on my belt carabiner. The South Rim Trail hooks into the Ribbon Lake Loop and the Loop will take us back to the parking lot where we began, a six-mile round trip. An ominous sign posts backcountry hazards: *uneven footing, drop offs, wildlife encounters.* What it doesn't say is that we'll also pass through a geothermal area and that we mustn't venture off the trail. An errant footstep can break through thin ground cover over hot thermal hazards. Hot feet may be the least of your problems.

At Point Sublime, the trail turns away from the canyon. We're on the plateau above the canyon and the going is mostly flat. Except for the wetlands around Ribbon Lake, Lily Pad Lake, and Clear Lake, we're passing through lodgepole pine forest or dry open meadow. Steam rises from hot springs near Clear Lake. We encounter no wildlife: no bears, no butterflies, no bird calls. We step around piles of bison dung. (Is it really bison dung? How do we tell the difference between that and bear scat?) We're hiking at a fast pace. The silence, the sulfur-scented steam, the potential for grizzly encounters unnerve us. When we reach the parking lot trailhead, Susan points out a red-lettered sign: "BEAR ATTACK: are you prepared to avoid one?"

Bears aren't the only disquieting Yellowstone hazard. The Park Service spends a lot of time reassuring visitors that they don't face imminent volcanic destruction, at the same time affirming that Yellowstone geology is dynamic. Earthquake swarms (hundreds per day). Swelling and collapsing of the caldera (a ten-inch swelling from 2004 to 2009, followed by a similar subsidence in 2010). Fluctuations in geyser and hot spring behavior. Even Yellowstone Lake recently tipping down to the south. These events occur over many square miles and would precede an eruption only if they occurred at the same place, at the same time, which, for now, they haven't. The park hosts a comprehensive volcano-monitoring program comprising twenty-six seismic arrays, sixteen GPS sensors, and eleven steam-gauging stations plus temperature, chemistry, and gas monitoring at selected geothermal sites. In addition, recent seismic studies indicate that Yellowstone's magma reservoir is further underground than originally believed and that the plume tilts downward and to the southwest, partially under the Eastern Snake River Plain. (This is good news for Yellowstone but not so good for the Snake River Plain, where future eruptions in the vicinity of Idaho's Craters of the Moon National Monument are likely.) The bottom line: no eruptions appear to be imminent and even if it happened, the most likely form would be basalt flows like those that occurred seventy thousand years ago, not the supervolcano eruptions of the more distant past. Will Yellowstone erupt again? Yes. Will it be a supervolcano eruption? Nobody knows. Will it erupt soon? Not likely.

Tomorrow, we depart. We haven't seen a geyser yet and a visit to Yellowstone without seeing a geyser wouldn't be a Yellowstone visit. Of the hydrothermal features in the park—hot springs, mud pots, fumaroles, travertine terraces, and geysers—geysers are the most dramatic.

A geyser is essentially a hot spring with special plumbing. It needs a heat source and plenty of water. Heat comes from the magma underneath the Yellowstone Volcano. Water comes from Yellowstone's plentiful snow and rain. The plumbing comes from the character of Yellowstone's volcanic rhyolite rock: It cracks easily, providing a constriction in the plumbing, usually near the surface.

Here's how the cycle works. Water sinks into the ground through cracks in the rock. As it does, the pressure increases due to the weight of the overlying water. The temperature increases due to the magma chamber's heat. The sinking water becomes superheated, meaning that its temperature rises well above the boiling point at the surface (199°F at Yellowstone's altitude) but it doesn't boil because of the water-weight-induced pressure. Because this superheated water weighs less than the same volume of cold water, it begins to rise. As it approaches the surface, pressure decreases and some of the water begins to boil. At a critical point, the boiling bubbles lift the water column, water flows out of the geyser, pressure further decreases, the superheated water violently flashes to steam, and steam and hot water jet out. Presto! You have a geyser eruption. Eventually the eruption stops because the geyser's water reservoir can't refill as fast as the eruption is emptying it. There are only a thousand geysers in the world. Over half are in Yellowstone.

Susan and I decide to drive south to the West Thumb Geyser Basin, not the closest to Canyon Campground but one I haven't visited since I was a kid. To get there, we must drive through Hayden Valley, where we encounter a major "bear jam," not for bears this time but for a herd of bison lumbering down the highway. We defer a stop at the Mud Volcano. There are too many tourists and no parking spaces.[2] We skirt Yellowstone Lake, bypassing Fishing Village, Lake Village, and Bridge Bay, the last already closed for the season. We reach West Thumb later than planned. We're also low on gas. Prior to the 1980s there was a campground here, a store, and a gas station. The Park Service moved these to Grant Village so as to preserve the natural beauty of the basin.

What makes West Thumb Geyser Basin unique is its proximity to Yellowstone Lake. Several geysers erupt directly from the lake. We park in an enormous lot and set out to walk the loop trail through the basin. Perhaps it's the tour buses, perhaps the shop selling T-shirts and tourist trinkets at the head of the trail, or how the grass between the parking rows has been beaten down by thousands of tourist feet, or the tourists taking selfies in front of the blue-green hot-spring pools—it's not even particularly crowded—but there's an aura of exhaustion about the place, as if it's seen too many tourists, as if it needs its winter's rest. A half-dozen small lakeside geysers sputter listlessly along the shore.

We gas up in Grant Village, a functional if charmless place. Then we retrace our route back to Canyon Campground.

But I don't want to remember Yellowstone by the West Thumb parking lot or the Grant Village gas station.

I want to remember Yellowstone in winter.

A Yellowstone Winter Coda

It's February, seven months before our current Bambi trip. Susan and I are standing in a blizzard at Biscuit Basin in Yellowstone National Park. The ruby-red Bombardier snowcoach that brought us here has just vanished into blowing snow. Ours, named "DeLacy," was a squat, minivan-like vehicle with skis in front and caterpillar treads in back. It was the youngest Bombardier in the Yellowstone Alpenguides fleet, although youth is relative—DeLacy was built in 1981 and once served communities in Canada's Hudson Bay. We're supposed to meet our driver and fellow passengers at Old Faithful Geyser Basin, six miles away, in about three hours—if we can find the ski trailhead.

"Do you have a cell phone?" our driver asked as he handed us our skis and blithely gestured in the direction of where the trail was supposed to begin. "If you run into trouble, call on your cell."

Cell phone, yes. Charged battery, no. Who knew there was cell coverage in Yellowstone?

We're suited up in layered skier attire. We have "classic" Nordic skis, a rucksack with the bare-bones backcountry "Ten Essentials," box lunches provided by Alpenguides, and a map (more like a brochure) of where the trail is supposed to be. Susan and I are veteran Nordic skiers, but we usually ski groomed trails. This time we can't even find where the trail begins. After a panicked search, we finally find a sign reading "Upper Geyser Basin" with an arrow pointing right. Upper Geyser Basin is what we want. A pair of ski tracks disappear into a copse of lodgepole pines. The tracks are rapidly filling with falling snow. Is it too early to call for rescue? Will our cell phone even work?

We're here because of a dream, and the dream is to see Yellowstone in winter. Aside from Mammoth Hot Springs, where automobiles are allowed year-round, there are only three winter ways to view the park: snowcoach (there are several providers), snowmobile (by permit or guided tour), or "human-powered" on skis, snowshoes, or hiking (guided or on your own). We've opted for a combination: snowcoach from West Yellowstone to Biscuit Basin and skis from Biscuit Basin to Old Faithful. We had expected other snowcoachers to ski with us. As it turned out, only we chose to ski. Maybe the blizzard deterred the others. The snow falls harder. The wind blows. It's now or never. We set off.

The trail is better marked than we initially feared. We spot ski tracks, tree slashes, even colored ribbons. There's no sound except for the whisking of our skis through the snow, the metronome beat of our hearts, the inhalations and exhalations of our breathing. The map signifies that we'll cross the Firehole River three times. The first won't occur until we're halfway to Old Faithful. Until we reach the Upper Geyser Trail, the trail will be flat. We begin to appreciate where we are: a pristine, primeval winter wilderness. White and gray. Snow softened. Sulfur-ponged. We pass a small hot spring. Steam rises from a blue pool. The trees open up. Now we can see steam from other geothermal phenomena along the river, whether fumaroles, hot springs, or geysers we can't tell. We cross the Firehole on a wooden bridge. In summer, it would be crowded with hikers. Today it carries only us.

We turn left onto a wooden boardwalk. We're in the geyser basin now and the boardwalk is here to keep us from skiing into a hot spring. A dark form looms through the falling snow. What's that? A bison! Two bison! Two tons of bison! Snow-coated. Breaths steaming. No more than thirty feet away. We've never been this close to such enormous wild animals. What do we do if a bison blocks the wooden walkway? Turn back? Ski off the boardwalk? The shaggy animals ignore us. Suddenly, from under the causeway, a gray creature shoots out, stops, turns, stares. It has blue eyes. A coyote or a wolf? The canine lopes off into the snow. Have we entered a winter paradise, one where bison and wolves lie down together?

Anxiety gives way to euphoria. We are dream-skiing through a winter Eden.

Of course, winter may not be Edenic for Yellowstone animals.

Today the temperature is close to zero. The wind chill is worse. Snow and ice cover the ground. Food must be hard to find. But these animals evolved in the Ice Ages. Are they not inured to winter? And isn't there something cleansing about winter? A time when older and weaker animals pass on? When elk and bison and bears gestate?

We're in the Upper Geyser Basin now, the site of Morning Glory Pool, Grotto Geyser, Giant Geyser, Oblong Geyser. All around us the sounds of muttering, spouting, steaming, bubbling, burbling, boiling. So much hot water drains into the Firehole that the river doesn't ice over. We pass by Beauty Pool, Grand Geyser, Turban Geyser, Sawmill Geyser.

The wind has abated. Snow flurries lightly. We have a rendezvous set at Old Faithful at two p.m. and we haven't lunched yet. We pick up the pace. For the first time, we meet other skiers.

"Where from?" they ask.

"Biscuit Basin," we answer with pride, the vanity of the trail.

We ascend a hill. The trail narrows and drops now to the river. For the first time, the skiing challenges our skills. We have to navigate a steep downhill turn. We cross the Firehole again. The Old Faithful Visitor Center rises before us. We ski around the famous and much-loved Old Faithful Geyser, for the moment quiescent, solitary, no tourists. We unstep our skis outside the visitor center. No food is permitted inside, so we lunch under the center veranda. We're chilled but ecstatic and our ecstasy wards off the chill. We enter the center just in time to watch Old Faithful erupt in spraying, steaming, streaming winter glory. When we ask a ranger about our coyote/wolf, she tells us that neither coyotes nor wolves have blue eyes. Our Bombardier driver comes over to the visitor center to find us. It's time to leave, he says. He was worried when we hadn't shown up at the cafeteria. We thank him for finding us and follow him back to DeLacy and a rumbling, rattling return to a less magical world.

I love Yellowstone. I love the idea that a place like this belongs to the people. I love it that the idea of a people's national park began here. I love knowing that Yellowstone is here even when I'm not—the geysers, the grizzlies, the mud volcanoes, the lake, the bison. But I worry about its fate. There are hazards: oilmen who want to drill for gas on its borders, ranchers who want to cull its grizzlies and wolves, mining companies who want to dig for gold in its watershed, animal-rights activists who protest when the Park Service tries to manage its bison herd, visitors, like Susan and me, who love the park but also threaten to love it to death. Still, in a bear jam, when you see folks snapping photos of wildlife, at a geyser outlook as tourists cheer its eruption, at Artist Point watching an octogenarian couple gaze with appreciation at the Lower Falls, you realize there's a symbiosis here between the park and people: something joyful, something we can't afford to lose, something that gives Yellowstone hope.

CHAPTER 10

SKY-HIGH WYOMING

What are men to rocks and mountains?

—JANE AUSTEN

THERE'S A BULL MOOSE loose in this campground and he's attacking campers' tents. We see clues: shreds of tent hanging out of the campground dumpster; no people in the campsite next door and all their possessions—stove, chairs, cooking gear—strangely staked down under a low tarp; and an agitated old man, a fellow camper, approaching us with his camera and babbling about tents and a moose (or was it an elk?).

We began our Wyoming loop here in Grand Teton National Park, at Colter Bay RV Campground, five days ago. Usually we avoid the RV-only campgrounds. You get electricity and water; you give up scenery, campsite size, and distance from neighbors. But we needed to fill the Bambi's water tanks and charge our batteries. At least Colter Bay had trees. And it was quiet. Everybody had electricity; nobody needed to run their generators.

The idea behind a Wyoming loop was to rectify our ignorance about the state. Friends of ours, full-time RVers, had urged us to visit the Wind River Mountains, among the most beautiful, they insisted, of all the Western mountains. Another friend, who grew up in

Wyoming, has proselytized us for thirty years to visit her home state. Now the loop is complete. We're taking a break and kicking back. We've been traveling Wyoming highways and the weather has been Old Testament.

The cranky-moose spot is the Gros Ventre Campground in the southeastern corner of Grand Teton National Park. Gros Ventre is sited on the east side of Jackson Hole Valley on the Gros Ventre River floodplain. The river meanders down one side of the campground. A sagebrush meadow and floodplain berm border the other. To the northwest, in the distance, the jagged Teton summits score the horizon: slate gray, mostly snow-free this mid-September season. Cottonwoods, a dusty late summer green, shade our campsite. The weather is mild, the sun out.

The campground concessionaire confirms a moose–tent attack. "A cranky old bull," she says.

"Not the first this season, I'll bet."

The woman laughs. "If you mean cranky customers, I suggested the same to another camper. He resented it." You can still see the moose, she adds, in the meadow between the campground and the Gros Ventre Road. And we do, an antlered bull lying in tall, dry meadow grass, glaring at us and the other tourists.

If there's an animal iconic of the Grand Tetons and Jackson Hole, the valley that foots the Tetons, for me it would be the moose, maybe because this is the place I saw my first one as a boy. Moose are the second-largest land animal in North America (only bison are larger). A bull moose may weigh as much as 1,200 pounds. Females, called "cows," are smaller, at around seven hundred pounds. They live solitary lives. The closest relationship a moose is likely to have is between a cow and her calf, who remain together until the cow births her next calf, which is about a year. During mating season males and females don't herd together—moose don't herd, period—but they do signal each other by grunts and snorting specific to each sex. Bulls will fight each other for the favors of a cow. Despite Bullwinkle, the warm-hearted moose in the *Rocky and His Friends* cartoons, moose aren't friendly. According to Wikipedia, they injure more people in attacks than any other wild animal. Wyoming's state animal is the bison but, in my opinion, the moose would be a better fit: solitary, irascible, ready to fight, no provocation required.

What damage could a bull moose do to an Airstream Bambi?

What makes the Tetons so unique is how they rise from the floor of a relatively flat lake-and-river-laced valley. There are no foothills. You cross a valley and then you face an abrupt, jagged wall. Here, ranchland lies adjacent to conifer-shadowed lakes and cottonwood-shaded river bottom in apparent peace. Cattle graze beside elk and bison wildlife refuges. A world-class ski resort adjoins a world-famous people's national park. A tourist town offers four-star dining, fake-cowboy-gear emporiums, a tourist-trap arch of elk antlers, and a second-home playground for movie stars, media moguls, Wall Street financiers, and ex-vice-presidents.

Are the Tetons typical of Wyoming mountainscapes?

Are the juxtapositions between park and natives harmonious?

Is it really Wyoming?

To all three questions, I'd answer, No.

Of the lower forty-eight states, Wyoming is the fourth largest, the third driest, the second highest, the least populated and least densely populated, and the most Republican (as measured by percentage of representatives in the state legislature). Despite its present ultra-macho pickup-truck-and-gun-rack persona, it was the first US territory or state to grant women the right to vote, this more than fifty years before the Nineteenth Amendment granted that right nationwide. It's not the windiest state (Kansas has that distinction) nor the least diverse (see Maine, Vermont, and New Hampshire), but it can be windy and it isn't diverse. It's the fourth coldest state year-round, although it often feels colder. I have driven across Wyoming when blowing snow and a setting sun made Interstate 80 look like it was a highway to heaven. I have driven Interstate 90 south of Sheridan in the fall and watched pronghorn antelope lift their tails and spring across the high Wyoming plains at sunset. At these moments, you may feel as I do—that Wyoming offers a rare, bleak, solemn kind of beauty. Not a softie's beauty. But I don't really know Wyoming.

At the beginning of our Wyoming loop, Susan and I ride our bikes from the campground/RV park down to the Colter Bay Visitor Center. This sits on the shore of Jackson Lake, the northernmost and largest of three glacial lakes that foot the Teton Range. The east shore is flat, geologically part of Jackson Hole. The west shore shoots up six thousand feet. That steep rise is what makes the Tetons famous. Why is it so steep? The answer is the Teton Fault. Two crustal blocks touch here. For the last twenty million years the earth's crust in this part of the Rocky Mountains has been stretching. The Teton block tilts up. The Jackson Hole block tips down. This uplift-and-dropdown process occurs in violent spurts with big magnitude 7.5 quakes. The rise after each quake may be as little as ten feet but geology plays a long game. We don't even see the full fault offset, which runs below the valley floor another twenty thousand plus feet. The process continues. Mountain building has a long time to run in the Tetons.

The other reason for the range's fame is the spire-like peaks called *horns* and *arêtes*.

What created these?

Two answers: Hard rock. And ice.

The Teton Range holds some of the oldest rock in North America, mostly gneiss and granite, formed over three billion years ago, when two continents collided and drove the surface rock deep below the earth's crust. Underground high temperatures and pressures, a process geologists call metamorphism—the word means "change of form"—physically and chemically altered the rock. Metamorphic rock tends to be hard. Other types of rock, like sandstone, overlaid the older rock. When the Teton Range began to uplift and tilt twenty million years ago, sandstone still capped the Teton summits and also formed the basement rock underlying Jackson Hole. The sandstone eroded faster. Then, only two million years ago, the Pleistocene ice ages began. The ice advanced down from the Yellowstone Plateau, scooping out Jackson Lake and Jenny Lake and bulldozing Jackson Hole's valley and leaving in its trail glacial debris and glacial moraine. The ice arrived and departed several times. At its maximums, two thousand feet of ice stood where the town of Jackson stands today. But this wasn't the only ice shaping the Tetons. Alpine glaciers in the Tetons themselves carved U-shaped glacial valleys, leaving behind knife-edged peaks between them, the cathedral-spire summits.[1]

The preservationist-versus-development battles that characterize the Mountain West have been fought over Jackson Lake. Originally a natural glacial lake, its level was artificially raised by the Jackson Lake Dam. The Snake River begins here and the dam is one of three on the Snake that provide irrigation water for Idaho's Snake River Plateau farmers. When the lake was added to the park, the groups struck a compromise. Idaho would get its water, the lake would be added to the park, and the dam would remain. Grand Teton National Park has made a number of these compromises. Cattlemen, for example, still graze their herds within the park boundaries. But compromise hasn't always been welcome. When John D. Rockefeller decided to preclude commercial development between Yellowstone and Grand Teton, he made his land purchases secretly in order to avoid local opposition. Like the mountain uplift and Ice-Age ice, preservation also moves in fits and starts.

We lock our bikes to a rack outside the visitor center and set out on a lakeside trail. As in Yellowstone, there are bear warning signs, both for grizzlies and black bears. We don't encounter either, but it does begin to rain. We head back to our bikes and our Bambi shelter. The weather has turned. Wind shakes the Bambi. Rain clatters on the aluminum.

Is this a premonition for our Wyoming loop?

The next morning, we leave Colter Bay and head south on US 26 through Moran and Moose Junction. We pass a long fence that borders the National Elk Reserve but we see no elk. After navigating Jackson's tourist traffic—we skip the elk-antler arch—we follow US 26 along the Snake River to its junction with US 189/US 191. Here we turn east. We're skirting the Gros Ventre Mountains, which lie east of Jackson Valley, and circling below the Wind Rivers, which are now north of us. The highway follows a narrow valley bordered by pine-clad hills until it reaches the small town of Bondurant. After that, the land opens up. There are no mountains in sight. We're actually traveling away from mountains. From Pinedale to Boulder to Farson, we enter a dry, rolling sagebrush plain. This is an area with cattle, the first oil and gas drilling rigs, barbed-wire fences. At Farson, we turn north on Wyoming 28. This takes us over South Pass, a key route in the Euro-American settling of the West: The Mormon, Oregon, and California wagon trails crossed the Rocky Mountains here, a surprisingly easy and distinctly unmountainous passage. South Pass is actually two passes; without the road signs you might not notice either. For Susan and me, veterans of California, Colorado, and Cascade Mountain passes, treeless rolling plains don't seem like a mountain pass. West of us, the sky darkens. The wind pipes up. Sideways gusts swerve our rig. I grip the MDX steering wheel. Suddenly the driver's-side trailer mirror flies away, bouncing down the highway, shards of glass in its wake. Goodbye, mirror. Fortunately, we have a spare, a smaller mirror left over from a previous mirror loss. We pull off the highway at the South Pass Rest Area. A semi idles next to us. Its driver stands only a few feet from our car door, tinkering with his truck trailer. I retrieve the backup mirror from behind Susan's car seat. When I unzip its bag, a gust catches it and the empty bag kites up and over the highway. The truck driver and I watch it sail east across a sagebrush sea.

"That bag," the semi driver says, "is on its way to Nebraska."

Ahead of us are Lander, Fort Washakie, Crowheart, Dubois, and the Wind River Indian Reservation. Behind us are three National Forests—Bridger, Shoshone, and Teton—and five different highways:

US 89, 26, 189, and 191 and Wyoming 28. We've crossed the Continental Divide several times (the divide here, as in Nevada's Great Basin, enclosing a basin where rivers and streams flow to no ocean). We turn northwest on a new highway, US 287. The Wind River Mountains—the main reason we're here—remain MIA, at least to our eyes. Maybe it's the weather. Maybe it's our reluctance to venture off the main highways. Maybe we're not there yet. Maybe sometimes a road trip doesn't lead you to where you want to go. It's getting late. We decide to find a place to stay in Lander.

Lander turns out to be a pleasant town that caters to backcountry tourists, dude ranch visitors, and students at the National Outdoor Leadership School. The school, founded in 1965 and popularly known as NOLS, is headquartered here.[2] The town also turns out to be the self-styled gateway to the Wind River Wilderness. (*Voilà!* Maybe we weren't there yet!) We decide to make our base in Lander and to venture into the Wind Rivers tomorrow. We find a small RV park, set up the Bambi, plug into water and electricity.

"Turn off the water tonight," the park manager warns. "It's going to get cold."

Cold? In mid-September? Despite the wind and the glowering clouds to the west, the sun is out and the temperature is in the mid-sixties. Cold seems unlikely. We find a nice restaurant in Lander with gourmet cuisine. We return to Bambi, watch a little TV, fall asleep, and wake up the next morning to a beautiful, blowing September blizzard.

The blizzard has caught even Wyoming off guard. US 26 to the west of Lander, the route back to Jackson Hole, is closed; the Wyoming Department of Transportation is scrambling to plow it open. The Forest Service roads into the Wind River Wilderness are impassable—nobody's plowing them. Weather, as it turns out, can change fast in sky-high Wyoming. We decide to hang loose in Lander. But we'll try to visit Sinks Canyon State Park, only a few miles south of town. There is a "good road up to the park," the RV Park manager assures us. "Should be okay in the snow."

We set out for Sinks Canyon. Snow muffles the sound of our wheels on the highway. We climb into a canyon bordered by pink-hued cliffs trimmed in snow. Pines and Douglas fir top the cliff walls. A small river, the Popo Agie (pronounced po'-PO-shuh) runs down the canyon floor, its banks overshadowed by snow-draped aspens and conifers. We pass a small campground where a few hardy RVers are hunkered down. We stop at an outdoor visitor center. Signs explain that the park is called "Sinks Canyon" because up-canyon, the Popo Agie vanishes into a limestone cavern called "The Sinks." Down-canyon, a quarter mile later, it reappears in a trout-filled pool called "The Rise." It takes two hours for the river to get from The Sinks to The Rise, and the flow there is larger than the flow that entered. Nobody knows what path the river follows but geologists speculate that groundwater from under the limestone formation augments its flow. Susan and I stand on the edge of The Rise. Hundreds of trout circle the clear pool. Snowflakes dapple its surface We have the place to ourselves. Pool. Trout. Snow. A moment, unexpected, unrepeatable, magical. This is why we Bambi-travel.

The next day the blizzard has passed. Sun out, highway cleared, we continue west, through Fort Washakie (established, we subsequently learn, to protect the Wind River Indian Reservation from the white settlers rather than the settlers from the Indians). After US 287 joins US 26, we climb through rolling foothills white with new snow. To the north, we see the Absaroka Mountains, to the south, the Wind Rivers. From our distance, both ranges, low on the horizon, appear less dramatic than the Tetons. Close up, we suspect, we'd be more impressed.[3] We pass through Crowheart—a reservation town—and then Dubois, where Butch Cassidy of *Butch Cassidy and the Sundance Kid* fame once owned a ranch. We cross the Continental Divide at Togwotee Pass. From there, we see the Tetons again: abrupt, jagged, gray, lacking in September snow. In another hour, we'll enter Gros

Ventre Campground, where an angry bull moose will have just attacked a tent.[4]

What else have we learned about Wyoming? Less than we expected. Our loop was too brief, the mountains too distant, the natives unspoken to. This illustrates yet another dimension of Bambi travel: we learn by travel where we ought to travel more.

Two Medicine
Scenic Cruises
Departures:
9:00 10:30, 1:00, 3:00 5:00

Ticket Sales And
Rental Information
At Ticket Office

CHAPTER 11

FOLLOWING THE ICE

US 2, Coulee City, Washington, to East Glacier, Montana

> Some say the world will end in fire,
> Some say in ice.
>
> —ROBERT FROST

IT'S MID-JUNE. SUSAN AND I are cycling through Steamboat Rock State Park in north central Washington State. We're also cycling across the floor of the Grand Coulee, the largest coulee canyon carved by the Pleistocene Ice-Age floods that swept across Montana, Idaho, Washington, and Oregon, the same floods that carved the Columbia River Gorge.

Steamboat Rock is much beloved by Washingtonians. Its hot, dry summers offer respite from the wetter western third of the state, and Banks Lake, which surrounds the park, provides warm-water swimming, water-skiing, and wakeboarding. The lake also serves as the primary distribution reservoir for the Columbia Basin Project, which feeds Columbia River water to vast portions of high-desert Washington and has, in my lifetime, made the Columbia Plateau bloom. On this Bambi trip, Susan and I will follow a highway, US 2, from here through the northern tops of Washington, Idaho, and Montana until we reach Glacier National Park. What we'll see along the way will reveal a story of Ice Age cataclysm and geological detective work, environmental catastrophe, a memory journey into my road-trip past, and grizzly bear tales.

Except for Steamboat Rock, the floor of the Grand Coulee is wide and flat. Basalt walls tower over us, in some places by over 1,300 feet. Because the rock is basalt, and because basalt often forms vertical multisided columns, the walls resemble Lhasa temples or castellated battlements. Black talus rock skirts the base of the cliffs. There's a surprising amount of cliff color: green, gray, rust, tangerine.

The word coulee comes from the Canadian French *coulée*, meaning "to flow." The French-Canadian trappers who applied the word to these canyons had no idea how apt it was. Picture a wall of water a thousand feet high sweeping toward you at sixty miles per hour, in its van, mud, icebergs, logs, dead bison and mastodons, boulders. Because the first peoples had just arrived and forty to fifty floods would occur over the next 2,500 years, at least every other generation would have witnessed a flood. What tales they must have told!

Paradoxically, twentieth-century geologists were slow to accept the floods. It took an irascible field geologist named J. Harlan Bretz to recognize the many Columbia Plateau landforms as evidence of massive floods. But where had the water come from? It took another geologist, J. T. Pardee, to identify the source by noting the strandline shelves—those created along lakeshore beaches—that had been left behind by a giant Ice Age lake north of what is now Missoula, Montana. The lake formed, Pardee theorized, when a lobe of the Cordilleran Ice Sheet, which then covered the western half of Canada, blocked the Clark Fork River, damming the lake back into Montana valleys. When the ice dam catastrophically collapsed, floods swept out of Montana, across Idaho, and into Washington and Oregon.[1] In the late 1920s, Pardee reported to Bretz that he might have found the water source for the floods, but Pardee didn't publicly support Bretz's theories until the 1940s, possibly because of the attitude of his superiors in the Geodetic Survey. For over a quarter of a century, geologists, many of whom were attached to Ivy League institutions and thus unfamiliar with Columbia Basin geology, argued against the notion of catastrophic floods. The dominant philosophy of geology at the time was known as *Uniformitarianism*: the idea that geologic processes happen slowly and operate the same way, whether now or in the distant past. This allowed geologists to interpret rocks from the past in light of processes they observed in their own time—a valid and useful methodology. But there was no "today" model for Bretz's hypothesis of catastrophic floods. Bretz, for his part, relished the controversy and enjoyed baiting his desk-bound East Coast colleagues. Finally, in 1979, the geological establishment acknowledged Bretz's work by awarding him the prestigious Penrose Medal—one of the highest honors in geology.

Susan and I head east on US 2. Sprinklers stretch across pale green fields like colossal aluminum centipedes. The water comes from the Columbia River via the Columbia Basin Project. Without the Project, these fields would be high and dry, a sagebrush desert. In the small towns along US 2—Coulee City, Hartline, Almira, Wilbur, Creston—we pass what seems to be an incongruous collection of fishing-boat and jet-ski dealerships, with no river or lake in sight. But Lake Roosevelt, dammed behind Grand Coulee Dam and the largest lake in Washington State, lies only a few miles north, a thousand feet lower from where we are. As we approach Spokane, islands of ponderosa pine begin to appear. Wheat gives way to cattle. Strip malls scar the highway approaching Fairchild Air Force Base, once the home of ready-alert Cold War B-52 bombers, some of which were always airborne to deter atomic war with the Soviet Union.

At Spokane, US 2 turns north. (The city's name, properly pronounced SPO-can, is often said incorrectly by outsiders, much to the annoyance of Spokane locals.) In the early 1970s, it hosted a World's Fair but it seems to have stood still since then. The city has a pleasant mid-twentieth-century feel, with riverside parks, navigable traffic, and modest brick and stone buildings. North of town, we pass through a series of shopping centers, suburban housing, and malls before entering open pastureland and more stands of ponderosa pine.

Soon US 2 turns east into Idaho. Pine gives way to hemlock, cedar, and Douglas fir. From here across Idaho's "Pan Handle," the narrow top end of the state, and into northwestern Montana, we'll transit what the Glacier National Park website calls "a finger of maritime climate," more like coastal Washington and Oregon than high-desert plateau or Rocky Mountain summits. We will also cross a geological boundary. Until now, we've been traveling over what were once island arcs in the Pacific Ocean until, about sixty million years ago, they "docked." Now, having reached the original plate boundary, we ascend the first escarpments of the Rocky Mountains, uplifted as the Farallon Plate carrying those island arcs slid under the ancient North American Plate.

Here more recent geological drama has also taken place.

At the town of Sandpoint, Idaho, at the north end of Lake Pend Oreille, we're close to where, some fifteen thousand years ago, almost yesterday in geological time, the Purcell Lobe of the Cordilleran Ice sheet dammed the Clark Fork River. The resulting lake, Glacial Lake Missoula, was as large as Lake Ontario and Lake Erie combined and was the lake that flooded the Columbia Plateau. At present-day Missoula in Montana, the lake was two thousand feet deep. It extended from Sandpoint over two hundred miles east and south. When the ice dam collapsed, and this likely occurred a dozen times, the ensuing flood exceeded the combined flow of all the rivers of the world. The Clark Fork River valley shows the effects of this by its width compared to other river valleys: The entire lake drained in two or three days and the surging water hollowed it out.

Several years ago, Susan and I followed the Clark Fork all the way to Missoula. We crossed a valley scoured by huge ripple marks, part of the evidence for the floodwaters that Pardee observed, telling of the volume and speed of the flood. This trip, however, we continue along US 2, which, after Sandpoint, turns almost due north to Bonners Ferry. The Purcell Ice Lobe descended through this valley.

At Bonners Ferry, we turn east again. From here we follow the Kootenai River. The Kootenai (spelled Kootenay in Canada) loops south from British Columbia into Montana and flows west into Idaho and then north again into British Columbia, where it ultimately merges with the Columbia. The river runs fast at the Idaho–Montana state line, even though it's dammed up- and downriver. Upriver, straddling the border, is a reservoir, curiously named Koocanusa, its name derived from CAN and USA cobbled to the KOO in Kootenai. As we cross into Montana, we're at the lowest elevation in the state.

The mountains here, on both sides of US 2, have the same Precambrian heritage, but they don't look much like each other. The reason is Ice-Age ice. To the north, in the Purcells, all but the highest peaks were graded and rounded under the Cordilleran Ice Sheet. The canyon in between, meanwhile, where the Kootenai now flows, was glacier or glacial lake. To the south, the Cabinet Mountains were never overridden by the continental ice but were shaped by their own alpine glacier system, leaving behind cirques and ragged peaks, the same type of horns and arêtes we saw in the Tetons.

We find a lovely Forest Service campground at the junction of the Kootenai and Yaak Rivers. The campsite sits on the river bank, shaded by giant cedars and hemlocks. There's a metal box-shaped bear locker for tent-campers to store food. I light our portable barbeque. We dine outside. Sunlight filters through the trees. The river plays an open-air melody. After dinner, I build a fire and stay awake long after Susan, assembling and reassembling the burning wood with a stick until all that's left are its red embers. I've never been here before but the place, especially the Yaak Valley, seems familiar to me. One of my favorite authors, Rick Bass, sets his short stories here. My favorite is "Fire," a tale of not quite unrequited love: A woman runner comes to the Yaak to train. A valley resident, male and unattached, cycles with her, armed with a pistol to ward off bears. It begins this way:

Some years the heat comes in April. There is always wind in April, but with luck there is warmth too. When the wind is from the south, the fields turn dry and everyone in the valley moves his seedlings outdoors. Root crops are what do best up here. The soil is rich from all the many fires, and potatoes from the valley taste like candy. Carrots pull free of the dark earth and taste like crisp sun. Strawberries also do well if they're kept watered.[2]

As I tend our campfire, I speculate that there are really three Montanas. The Rick Bass Montana would be the western third of the state: forested, mountainous, wet, temperate, more like the Pacific Northwest coast. After the Continental Divide, a second Montana begins, what I'd call "Rocky Mountain Montana," with the Continental Divide's eastern slope, the state capital, Helena, the mining capital, Butte, and Bozeman, the site of Montana State University and gateway to Yellowstone National Park, plus world-class trout streams, aspens, Douglas fir, high meadows, and a web of mountain ranges. Summers are hot there, winters harsh. After Bozeman, the easternmost and largest part of Montana, is what I'd call "Prairie Montana." Here, rolling high plains run all the way to the Dakotas, ranchers run thousands of cattle, farmers plow vast fields of wheat, and General Custer met his end at the Little Big Horn.

As we proceed east on US 2, a winding two-lane, frost-heaved rollercoaster of a road, I feel as if I've traveled back in time. In October 1955, my parents drove my brothers and me from Seattle to their hometown of New Kensington, near Pittsburgh, Pennsylvania. US 10 was then a two-lane rollercoaster road just like US 2 is today. My father had only two weeks' vacation. Air travel was expensive, especially for five of us, so he and Mother decided to drive what they said would be "straight through." "Straight through" meant no overnight stops and restaurant dinners and breakfasts, an uncommon luxury for my brothers and me. We set out with a mattress in the back of our Ford station wagon. Dad and Mom took turns driving while the other slept. My brothers and I tucked ourselves anywhere we fit. There were no seatbelts then. No superhighways, either, at least not until you reached Chicago, and even those were under construction. We followed US 10 across Washington, Idaho, Montana, North Dakota, Minnesota, Wisconsin, and Illinois. The journey took three days. I have marvelous memories of that trip: Montana mountains. Dakota prairies. The Missouri River. Combines reaping golden wheat. The smell of oatmeal, maple syrup, and frying bacon in the bus stations and truck stops where we stopped for breakfast. Coal smoke and toll turnpikes in Pennsylvania. Howard Johnson's ice cream in Pittsburgh. The breadth and plentitude of our continent-spanning country unfolded outside our station wagon windows, creating an appreciation for America's size and beauty that remains with me still. So here we are, Susan and I, on US 2, raveling memories from a traveling past.

We pass through Libby, a onetime logging and mining town trying to make the transition to tourism. Unfortunately, an environmental disaster darkens its history. Vermiculite was mined here, a material used in a variety of products from brake linings to industrial furnaces, not in itself hazardous. But the vermiculite mined in Libby *was* hazardous: asbestos tainted it, and asbestos causes asbestosis—a lung disease that leads to cancer and other illnesses. Because the mine byproducts were used to construct the town's homes and businesses, as many as 10 percent of Libby's inhabitants have died from asbestos-related disease. W. R. Grace & Co., which owned the mine, attempted to siphon assets worth billions of dollars to another Grace subsidiary before declaring the mining operation bankrupt. Government attorneys blew the whistle on that scam. Grace was forced to return the assets to the bankrupt subsidiary sufficient to cover the lawsuits. Victims have special eligibility for Medicare. The town is a Superfund site. The cleanup is ongoing.

We pass a series of lakes that owe their origin to glacial ice: Loon, Horseshoe, Crystal, and Thompson. As we travel east, the land opens up—the valleys grow wider, the mountains more distant. By lunchtime, we're in Kalispell, on the Flathead River, once the bottom of Glacial Lake Missoula. A hundred thousand people live here, the largest urban concentration in northwest Montana and the heart of Montana tourism: Glacier National Park, two major ski areas, and Flathead Lake are nearby. We lunch at a Mexican restaurant and head up to the park. Our destination is the large campground at Apgar, adjacent to Lake McDonald. We haven't made a reservation, presuming that in early June an early arrival will suffice. Our presumption is valid, but just barely. By late afternoon, Apgar is full.

We plan to explore the Lake McDonald side of the park and then make a day trip over Logan Pass on the Going-to-the-Sun Road. A dozen years ago, Susan, our son John, and I drove up the east side of the pass. Snowmelt water showered the highway; we felt as though we were driving inside a silver waterfall. The Going-to-the Sun is famous, not just for its scenery, but also for its elegant design, a single switchback that rises from the Lake McDonald side of the park to Logan Pass across the face of a mountain cliff. Its construction forged a partnership between the Park Service and what was then the Federal Bureau of Public Highways, birthing a style that came to characterize National Parks highways: roads built not just to get from one place to another but to reveal scenery and to preserve the character of place with stone-faced bridges, rustic guardrails, and even tunnels with mountain-view windows. The Going-to-the-Sun opened in 1933 to much acclaim. My parents drove the highway in 1950, towing our Papagayo trailer up and over the pass. Later, because of narrow lanes, low tunnel heights, and greater park traffic, the Park Service banned trailers and large RVs from the Going-to-the-Sun.

Our day's plan doesn't survive our arrival at Apgar. Signs at the park entrance declare the road closed. Late spring snow still blocks the pass. Fortunately, we've brought our mountain bikes and the park newspaper shows a number of biking trails in the Apgar/Lake McDonald area.

A bike-and-Bambi digression may be in order. Readers with no cycling interest may want to skip this section.

Bikes augment our Bambi-travel experience. In addition to giving us healthy exercise, they let us explore at a slower pace than driving but cover more territory in less time than we could on foot. Plus, more and more bike-only trails exist in parks and cities. But you have to figure out how to carry them.

We watched so many bikes bobbing up and down on the back ends of so many trailers that we decided to forgo a trailer-mounted rack. Instead, we mount our bikes on the roof of the MDX. Getting them up there, however, demands preparation, some physical strength, and a lot of balance. The bikes mount upright into two separate Yakima racks that run fore and aft on roof-rack crossbars. On our car, the handlebars face rearward so that the racks clear the MDX's tailgate and moon roof. Two hoops hinge at the front-wheel end of each rack—a large one that adjusts for the diameter of the wheel and a small one that braces the wheel back. To load the bikes, I flip up the rack's large hoop from its pivot. The small hoop lies flat. Then I open the MDX rear door, grip the bike by its frame, and hoist it up, laying its wheels across the rack's base, a grooved track just wide enough to fit the bike's tires. Yakima calls it "the forks." Holding the bike steady, I step up onto the open car-door sill. This is the dicey part because I need to balance there while gently tipping the bike to a vertical position. One slip and the pedals will scratch the car or I'll fall backward. I roll the bike rearward until the front wheel rests securely in the large hoop. Meanwhile, the small loop levers up, bracing the front wheel from behind. I spin the hoop-tightener, a large red knob at the base of the large hoop, until the wheel is firmly gripped. The bike's rear wheel already sits in the wheeltray. I guide the wheeltray strap through the bike spokes until it snaps into the strap buckle; then ratchet the buckle tab up and down until the strap is tight. The last step is to weave the antitheft cable through the bike frame and lock it in place.[3] On several occasions, passersby have offered to help, alarmed, I suppose, at the sight of a white-haired geezer like me balancing a bike on a car rooftop. Once, early in our Bambi travels, I forgot to use the hoop-tightener. One bike keeled over and off the rack—a bad outcome leading to a traffic jam and scratched rooftop paint. Installed correctly, however, we've driven thousands of miles with this type of rack, taking care to always remember the bikes are up there, especially if we're unhooked from Bambi. Susan attaches a sticky note to the car windshield that reads "Bikes on top!"

Tomorrow we'll bike. Today we'll explore by car.

We follow a two-lane highway along the Lake McDonald shore. Glimpses of the lake flash between stands of conifer trees. McDonald is a glacial lake and we're driving over a lateral moraine plowed to the side by a glacier's advance and retreat. A paradox of Glacier National Park is its dearth of glaciers. The most recent serious glaciation was during the Little Ice Age, a several-hundred-year span from about 1300 to 1850 CE, when the climate cooled. Scientists disagree why this happened, and many insist that the name is a misnomer for what may have been a local or hemispheric rather than global phenomenon. In any event, even the Little Ice Age didn't do the major work. The truly serious glaciation that carved the landscape of Glacier National Park and led to Glacial Lake Missoula and the Columbia Plateau floods occurred during the Pleistocene Ice Age, which ended twelve thousand years ago. The result is the classic glacial landscape we see in Glacier today: large U-shaped valleys; steep walls; long, deep lakes; hanging valleys; arête and horned peaks; glacial cirques and tarns; chains of paternoster lakes (so named for their resemblance to rosary beads) formed as glaciers retreated up valleys; and terminal and lateral moraines layered with boulders, cobbles, pebbles, gravel, sand, silt, and clay that form the ends and edges of glacial valleys and lakes.

We stop at the Lake McDonald Lodge. The lodge, built in 1913, is the oldest accommodation in the park, a collection of rustic hotel rooms and cottages in the National Park style, in addition to more mundane 1960s and 1970s motel rooms. The rustic portion of the lodge fronts the lake. Hanging-plant baskets dangle from the eaves. A glittering creek runs between the lodge and cabins. Guests can sit and view the lake from the lakeside deck and picture early twentieth-century visitors arriving by excursion boat: men in wide-brimmed hats, women in long dresses, the bellmen bearing leather trunks and suitcases.

At the hotel entrance, we spot the iconic "Red Buses" boarding passengers. I remember these from visiting Glacier as a boy (or maybe I remember them from Mount Rainier National Park, which had similar vehicles). The current Glacier fleet dates from the 1930s. The buses are elongated, like stretch limousines, although taller and with four rows of seats, doors on each row, an added feature being a roll-back top, perfect for mountain viewing. A bus seats seventeen people, four in each row and one adjacent to the driver. "Snug quarters," the Red Bus website warns. The drivers are called "jammers" for the sound the buses' manual transmissions made before the present-day automatic transmissions were installed. From 1914 through the 1970s, jammers were college-aged men, mostly pre-law or pre-med students. Female riders were discouraged from sitting next to the jammers for fear of distracting them while they navigated the Going-to-the-Sun. On this day, with the Going-to-the-Sun closed by snow, the buses are running a reduced schedule on revised routes. Still, every bus we see is full.

We dine at the lodge restaurant. The menu features rainbow trout, elk steaks, and bison burgers. When we return to our campsite, a light rain has begun to fall. We set up our open-sided "sunshade/ rainfly" tent, just big enough to cover the campsite picnic table. I build a campfire. We sit and poke sticks into the fire until the late evening light turns into nighttime darkness. A peacefulness attends us, but maybe you have to camp as a kid to appreciate it.[4]

The next day dawns gray with clouds and rain forecast. We decide to focus our bike ride on the area between the Apgar and Fish Creek Campgrounds, a short jaunt from which we can retreat if the weather worsens. From Apgar to the Lake McDonald Visitor Center, the bike paths are paved. We cross a short highway bridge over Fish Creek. From the bridge, we follow an unpaved road until we reach a mountain bike trail. Here, a sign reads "Entering Grizzly Country." It recommends that we carry bear spray. Ours is back in the car. We're tempted to proceed. The distance from here to Fish Creek is short. But the trailside forest is so dense it could easily hide a bear. We retreat back to the paved highway. Even so, we nervously glance from side to side as we climb and then descend into Fish Creek Campground. Only a few campers are here. Later in the summer, reservations will be mandatory. After we return to Apgar, the weather threatens rain, so we cancel another extended bike ride. Instead, we drive to where gates close the Going-to-the-Sun. There's a picnic area here jammed with parked cars. This is a popular trailhead, with short hikes to river viewing sites as well as longer trails into the backcountry. We opt for a short, serene trail through a cedar rain forest.

Because it straddles the Continental Divide and because three watersheds meet in the park, Glacier's creeks and rivers flow to three oceans: the Pacific, the Atlantic, and the Arctic. But Glacier's divide is more than a geological boundary: It's also an ecosystem boundary. As we follow US 2 eastward along the Flathead River, the forest lightens up: there are more aspens, fewer conifers, the underbrush still dense. US 2 tracks the Great Northern Railway over Marias Pass, also the route for Amtrak passenger trains. As we summit the pass, clouds obscure the taller peaks. On our return, two days from now, we will encounter hard rain. June isn't really summer in upper Montana. Still, we feel an openness here. Montana has more than the usual quota of state slogans and marketing monikers. "Oro y Plata" is the official motto but "Treasure State," "Bonanza State," "Mountain State," even "Headwaters State" have all been used and sometimes inscribed on Montana license plates. My favorite is "Big Sky Country." I suspect this slogan owes more to Prairie Montana than to Rocky Mountain or Rick Bass Montana, but it seems to me that the state's big, broad-shouldered mountains do reveal more sky.

At East Glacier, we leave US 2. The town is small, geared for tourists, and mostly empty this early in the season. We stop for lunch.

Huckleberry pie and ice cream are on the menu. Once this was the most visited locale in the park. Visitors arrived by train, spent a night in East Glacier, and then proceeded on horseback to Two Medicine Lake. Our friends, full-time RVers Marti and Joyce Waite, insist that the campground at Two Medicine is one of the best in the West. We follow the Two Medicine River on the frost-heaved but paved Montana SR 49. Thickets of aspens crowd the road. We pass Lower Two Medicine Lake, a reservoir, before reaching the park entrance.

The upper lake is situated in a valley below Sinopah and Rising Wolf Mountains. The log-sided former lakeside hotel has morphed into a camp store but the original boathouse still serves the crafts that ferry visitors across the lake. Both buildings and boats are listed in the National Register of Historic Places. We plan to hike the lake loop trail tomorrow, but now, mid-afternoon, we decide to set up camp.

Marti and Joyce are right. The campground is lovely. We find a site with perfect views, located amid a grove of Christmas tree–sized spruce. It's breezy and cool but the sun is out. Clouds scud around the knife-sharp peaks and across a big, blue Montana sky. As we sip our post-setup cocktails, a ranger knocks on the Bambi door. He invites us to the evening campfire talk.

"What topic?" we ask.

"The Grizzly Details," he replies. "About grizzly bears."

Grizzlies! That piques our interest.

After dinner, we traipse over to the campground amphitheater. National Park amphitheaters have a standard look and feel: backless benches arranged in a semicircle, sometimes a projection screen, occasionally a campfire. This one has no screen and no campfire, but it does have a view of Sinopah Mountain, gilded this evening by alpenglow. Perhaps twenty fellow campers have joined us.

The ranger is middle-aged, trim, his Park Service uniform green but without the trademark Smokey Bear hat. He begins with an audience participation question: "How do you tell a grizzly bear from a black bear?"

It's a sort of a trick question with the ranger trying to elicit a particular incorrect answer, but this crowd, especially one middle school–age boy, is into bears. "A grizzly has a hump behind its neck," the kid answers, "a 'dished' face, longer claws, and smaller ears."

"Anything else? How about color?" the ranger asks.

This is the trick part of the question.

"Not color," the kid answers. "Black bears and grizzlies have various colors."

"Absolutely right," says the ranger, although he seems a little disappointed that his bear-color ploy has fallen flat. He adds cheerfully, "We had a yearling grizzly male wander through the campground yesterday. Did anyone see him?"

A grizzly in our campground?

Nobody raises a hand.

I whisper to Susan, "Maybe we should reconsider our hike around the lake."

The ranger begins the familiar but not reassuring litany about how to act if you encounter a bear. We've heard the advice many times about making noise and carrying bear spray and not running away or trying to climb a tree. This ranger adds information about firearms, especially handguns, which, he says, are now permitted in National Parks due to the wisdom of the US Congress. These are not effective against grizzlies due to the thickness of the animals' skulls. "To shoot a grizzly with a handgun," the ranger says, "you'd need nerves of steel or have to be brain dead."

Susan and I don't have handguns or nerves of steel. The next day, in lieu of our planned lake circumnavigation, we opt for a boat ride across Two Medicine Lake and a group hike to a picturesque waterfall.

A week after we leave Glacier, a Forest Service ranger riding a mountain bike near Apgar, the same area where we rode our own, will round a trail corner and run into a grizzly bear. His companion will ride for help. When the help returns, the grizzly will be gone and the ranger dead, bones broken, body mauled.

We head south on US 93 in driving rain, through the Flathead Valley and beside Flathead Lake. It was here that J. T. Pardee observed the strandlines that evidenced the Ice Age lake that fed the Columbia Basin floods. At Missoula, the sun reappears. We are leaving the land of Ice Age ice.

CHAPTER 12

CALIFORNIA DREAMING

Highway 1, the Golden Gate to Malibu

What would we do without California?

—SUSAN COLE

IT'S MIDNIGHT, THIRTY MILES north of Santa Barbara. A tree has just crashed onto our Bambi. Somebody pounds on the door, yelling, "You guys okay?"

I leap from our bunk, take three steps to the door, and crack it open. Sparks gust from a nearby campfire, illuming the night in dancing light and shadows. Trees whip back and forth. Two campers—young men—are pulling the wind-downed tree away from Bambi's door. I push until the door opens. The tree lies between the Bambi and the car, its branches caught in our rack-mounted bikes. California has its hazards—earthquakes, tsunamis, wildfires, triple-digit heat. But hurricane-force winds? In April?

This campground, Refugio Beach State Park, lies just off US 101, here also called Highway 1, California 1, or the Cabrillo Highway.[1] Highway 1 spans the California coast. South from the Golden Gate— where we're traveling this Bambi trip—it's what people expect California to be: sea stacks, dunes, promontories reaching into the

Pacific, redwoods, eucalyptus, cypresses, palms, prickly-pear cactus, red poppies, green pampas grass, golden grassy mountains, oak savannas, vertiginous cliffs, chaparral cattle ranches, truck farms, oil platforms, lifeguard stations, beach bungalows, Malibu mansions, fishing piers, amusement parks, thousand-slip marinas, surfers, in-line roller skaters, bodybuilders, beach volleyballers, sea lions, pelicans, dolphins, and humpback and gray whales. Once you've departed San Francisco, few places until you reach Santa Barbara would be called wilderness. Few places would be called urban, either. For us Highway 1 is not so much an exploration as a homecoming. As soon as we enter California, we watch for our favorite touchstones. The first palm tree. The first eucalyptus. The first redwood. The first billboards advertising Trees of Mystery. One November, the first year of our Bambi travels, sun shining, seventy-five degrees, at a highway pull-off only a few miles north of where we are this night, we watched the Pacific surge through a coppery kelp forest. In Seattle, meanwhile, it was raining and forty degrees.

"What," Susan said, "would we do without California?"

On this April trip, a week before the tree fall, our Highway 1 journey begins at the Golden Gate Bridge. No Highway 1 touchstone is dearer to us. Cables drape in graceful arcs from twin towers that rise vermilion-orange to over seven hundred feet. Six lanes of traffic carry one hundred and twenty thousand cars each day. To the west lies the Pacific Ocean, to the east, the Bay and San Francisco with its bone-white buildings. The bridge was completed in 1937 and remained, until 1964, the longest and tallest bridge in the world. It has appeared in dozens of movies, ranging from *The Maltese Falcon* to *Dirty Harry* to *The Love Bug*. Susan and I have cycled across it several times, twice riding the ferry from Tiburon on the Marin County side to the San Francisco waterfront and then biking up and over the bridge. When I see the Golden Gate, I have the same feeling I have when I see the Statue of Liberty: optimism, dynamism, the promise of America.

There is no fog this morning. The lanes are narrow, so I pay extra attention to the Bambi's width. We exit the bridge and follow Highway 1 south into San Francisco. It soon merges into 14th Avenue in the Richmond District, then hooks west and south through Golden Gate Park, continues as 19th Avenue through the Sunset District, passes San Francisco State University and the Parkmerced Shopping Center, merges with Interstate 280, then dives oceanward into Pacifica. Until Pacifica, it's not trailer-friendly territory. We have to contend with residential cross streets, stop lights, narrow lanes, and, of course, heavy traffic. But it's civilized traffic, unlike the speeding, weaving, veering, accelerating, braking, shifting, shying California freeway traffic. Famously, in the 1960s the residents of the Sunset District rose up and stopped a plan to connect I-280 by a new freeway to the Golden Gate Bridge, an early case, perhaps the first, of a neighborhood saying "Enough freeway!" I have always liked the area's stucco row houses, pastel colors, and eucalyptus trees, as if you're in a Mediterranean seaside village rather than the foggy Northern California coast.

By contrast, Pacifica's charm is elusive. Proximity to the ocean may be charm, but it's a cold ocean and it feels precarious, as if one big wave might wash it all away. On a previous visit we stayed in the town's San Francisco RV Park. Its seaward side actually *has* washed away. A chain-link fence divides the park (mostly an asphalt parking lot with RV spaces delineated by white lines) from a bike path that runs atop a beachside cliff. The path is being undermined by the sea. Chunks of asphalt have already fallen to the beach below. Still, the park is popular with RVers. It is one of the few near San Francisco and the only one with ocean views (which are often, in any season, fog-shrouded).

From Pacifica south to Santa Cruz, Highway 1 runs west of the San Andreas Fault. It rests on the Pacific Plate, which is sailing toward Alaska and trying to drag the North American Plate with it. You can't quite escape the feeling that the landscape is ephemeral: One big quake and it might vanish. We're fond of this stretch of California coast. When we lived in the Bay Area, we'd drive over from Palo Alto in order to hang out on these beaches. In the early 1990s, they were crowded on warm weekends but not on weekdays. Now, year by year, we see more weekday traffic, more full parking lots, more lunchtime crowds. Ten thousand baby boomers retire each day. They're hitting the roads as elder surfers, beachcombers, birdwatchers. And the birds are plentiful: plovers, avocets, sandpipers, drumlins, tattlers, whimbrels, curlews all scurry stilt-legged before the breaking waves.

Not much farther south, in the town of Santa Cruz, we enter what I consider the northernmost California beach town. Like its sister beach towns, it's a conglomeration of tackiness and stylishness. From the car window we can see tattoo parlors, fast-food joints, T-shirt shops, a 1930s-era amusement park, fishing piers, bougainvillea-purpled parks, wind-bent cypress trees, statuesque palms, flower-bedecked bungalows, shoreside bike trails, surfers, and sunbathers. UC Santa Cruz is sited here, so it also has all the attributes of a university town—cheap apartments, coin-op laundromats, teriyaki takeouts, and down-and-out bars. Santa Cruz consistently scores as one of the happiest communities in California. But take care if you're a newbie. There's something called the "Santa Cruz Chill." Outsiders often aren't made welcome.

We park just south of the Santa Cruz Marina on a beachside street. A sidewalk lies outside our Bambi door, a stone wall, and beyond the wall, a sandy beach. We eat lunch and watch dolphins and whales. Every few minutes, pelicans soar over a gentle surf break, rising up and down, riding an invisible, airy rollercoaster. The upwelling of cold water from the ocean bottom is strong this year, and close to shore. This means bountiful baitfish for dolphins, whales, and pelicans, as well as other sea creatures. Every few minutes, a cyclist or pedestrian passes on the sidewalk. They all glance in the Bambi door. We greet them, raise a glass in toast. We know that if we linger too long, the police will ask us to leave. Santa Cruz guards its beach privacy. Lunchtime will be long enough today, a break as perfect as can be.

From Santa Cruz south, Highway 1 follows the scimitar crescent of Monterey Bay. We cross the Salinas River estuary and pass under the power plant at Moss Landing. The Salinas Valley borders the bay here. Not long after the last Ice Age, when the sea level rose. the valley flooded and became a shallow saltwater embayment. Eventually it filled due to deposition of gravel and sand from the surrounding mountains.[2] Now it's some of the most fertile agricultural land in California. Artichokes. Tomatoes. Garlic. Lettuce. Strawberries. Grapes. One morning, before sunrise, at Sunset State Beach Campground, we watched a tractor plow a strawberry field behind the campground, its headlights illuming the fog, a magical agrarian tableau.

This part of California is dear to me. I first visited Monterey in 1966 as a third-class midshipman aboard a US Navy guided missile frigate, the USS Gridley. We marched in the town's 4th of July parade, cheered by crowds as we tromped by. (Monterey was, and still is, a military-friendly town. Fort Ord, the Naval Post-Graduate School, and the Defense Languages Institute are located here.) I did my first California scuba dive off Cannery Row. In 1991, we took our toddler son to the newly opened Monterey Bay Aquarium, John's first aquarium visit, a place we've visited again and again. Susan observed on our most recent aquarium visit, "Have you noticed how happy the children seem?" Adults, too.

We often bike the trail that runs from Pacific Grove around Lovers' Point, past the aquarium, through Cannery Row, along the shore of the yacht harbor and Fisherman's Wharf, under a corridor of eucalyptus trees, then up into the dunes adjacent to the town of Marina, through historical Fort Ord, now a state park, and ending in the town of Seaside, an eighteen-mile round trip.

In fifty years of visits here, I watched Cannery Row evolve from falling-down abandoned buildings, surely familiar to author John Steinbeck, who set many of his novels and short stories here, to an upscale destination of luxury hotels and fine restaurants, without, I think, losing its essential character. We often lunch in Schooners in the Monterey Plaza Hotel. An outdoor patio overhangs the beach. Rollers break against old piers and old cannery buildings, the biggest breakers shooting thirty feet up the walls. Kayakers, paddleboarders, scuba divers play in the kelp. This trip, Susan and I launched our own inflatable kayaks and paddled out amid a cacophony of barking, odiferous sea lions, many bigger than our kayaks.[3]

Despite its tourism, Monterey is still a working town. Its next-door neighbor, Carmel-by-the-Sea, is not. Carmel's 17-Mile Drive is deservedly famous for its cypress-shaded points, rocky promontories, and cozy beaches. (A fee of $11.25 per vehicle, at this writing, is required to enter. It's worth it for at least one trip.) The city's golf courses host prestigious tournaments. Its longtime former mayor, Clint Eastwood, is famous. Its downtown typifies the West Coast boutique aura—small shops, little inns, no franchises. Like its former mayor, Carmel is understated, classic.

South from Carmel, we pass Point Lobos State Beach, gateway to the Big Sur. The name Big Sur generally applies to the seventy miles of coast that begins at Malpaso Creek in the Carmel Highlands and runs south to San Carpoforo Creek near San Simeon. When the first Spanish explorers came north, they bypassed Big Sur. The Santa Lucia Mountains fell so precipitously to the sea that the area seemed impassable. Eventually the Spanish referred to the region as *El Sur*, "The South," because it was south of the colonial capital at Monterey. For many years, it was left unsettled: As late as the early twentieth century grizzly bears still roamed Big Sur. When the Spanish built their California mission road, El Camino Real, they laid it to the east, on the other side of the coastal mountains. No road ran for Big Sur's full length until 1937. Because the mountains were fractured by canyons and seasonal stream beds, thirty-three bridges had to be built. Prisoners from San Quentin and Folsom were employed at thirty-five cents per day as unskilled laborers, as was a young aspiring writer who would win the Nobel Prize for Literature. His name? John Steinbeck. The Big Sur highway was also one of the first constructed less for economic and military reasons and more for the burgeoning population of California tourist-motorists who wanted a way to see this portion of the coast.

The road clings to the mountain face, sometimes a thousand feet above the sea, sometimes at sea level. Landslides frequently close it. In 2017, large slides at Mud Creek and the Pfeiffer Canyon Bridge closed the road for months. Much of the coast here consists of what's called the Franciscan Complex, an unstable mélange of sedimentary rock laid down at various ocean depths, in various locations, collected in ocean trenches along with chunks of seafloor basalt. Over millions of years this has uplifted, subducted, fractured, and scraped off to the terranes that underlie much of California.

Add a seasonal climate of dry summers and intense winter rains and you have a textbook environment for landslides. Originally the highway closed in winter. Now, if there are no slides, it's open year-round. Only a thousand full-time residents live in Big Sur. But three to four million people visit each year. They come for the dramatic seascape views. They come for a sense of what the California coast looked like pre-urbanization. They come to visit the small beaches (only a few of which are accessible) and to view the promontories and the sea-stack rocks and the surging Pacific Ocean. Mostly they drive through without staying. Only three hundred hotel/motel beds are available, and only a few state and Forest Service parks offer campsites. Most tourists try to make the trip from Southern California to San Francisco via Big Sur in one day. Traffic races around hairpin turns. Drivers hug the center of the road. Cars dodge in and out of scenic viewpoints. There are only a few gas stations and they sell the most expensive fuel in California.

Is the spectacle worth the hassle?

Yes.

Big Sur casts a spell. Maybe that's why artists and literary luminaries like Henry Miller, Jack Kerouac, Hunter S. Thompson, and Richard Brautigan settled down here. Maybe that's why several spiritual institutions are located here: the Tassajara Zen Mountain

Center, the Catholic New Camaldoli Monastery, the Esalen Institute, famous for popularizing Eastern philosophies and most recently for ad executive Don Draper's last scene in the *Mad Men* TV series where he visualizes the famous Coca-Cola spot as he sits in a lotus pose.[4]

Big Sur has cast its spell on Susan and me, too, one place in particular: a shelf of land between Highway 1 and the shore cliffs, a Forest Service campground shaded by pines, bright with yellow wildflowers, green with tall grasses, a blue Pacific to the west, where once, eighty years ago, the prisoners who built the Big Sur highway camped. It's called Kirk Creek, and for us, it typifies the best of Big Sur—beauty, silence, peace, sunshine.[5]

San Simeon, at the southern end of Big Sur, lies almost exactly halfway between San Francisco and Los Angeles. Here, the mountains fall back from the coast. The land opens up. Low, cypress-topped headlands enclose pebble-and-sand beaches. The land between the shore and the mountains is relatively flat. Once it was a wave-cut terrace, lifted up, perhaps suddenly, in an earthquake.[6] Now it's ranch country with cattle, horses, and rolling pasture. If you think you saw a zebra, you may have. William Randolph Hearst, the early-twentieth-century newspaper magnate, built an estate here: the Hearst Castle, now a California State Park. The zebra are descendants of the denizens of his private zoo.

Hearst admired Spanish-style architecture. He patterned the castle after a church he visited in Ronda, Spain. He was also a voracious collector, importing entire European rooms complete with ceilings. This was a challenge for his architect, Julia Morgan, especially because Hearst frequently changed his mind. In the 1920s and '30s, the estate was the site of elaborate parties attended by the entertainment and political elite, among them Charlie Chaplin, Cary Grant, the Marx Brothers, Charles Lindbergh, Greta Garbo, Joan Crawford, Clark Gable, James Stewart, Bob Hope, Calvin Coolidge, Franklin Roosevelt, Dolores del Río, and Winston Churchill. Guests stayed in separate bungalows but were expected to assemble in the castle for dinner. Hearst, who died in 1951, never completed his mansion. His family still uses the estate's original Victorian house, a proviso in the bequest to the State of California.[7] Nowadays, the State Park offers tours. Reservations are advised.

Below the visitor center, along the shore, lies the Hearst–San Simeon State Park. The park has two campgrounds, a number of trails, and several public beaches. From December to March lumbering male elephant seals, with their wobbly, drooping proboscises, battle for female harems on several beaches, a major tourist attraction with attendant traffic.

If the stretch from San Simeon to Morro Bay appears tranquil compared to Big Sur, this is an illusion. Earthquakes frequently rattle the area. The San Andreas Fault runs on the other side of the Santa Lucia Mountains, a strike/slip fault boundary between the Pacific and North American plates.[8] As Susan and I drive toward Morro Bay, we see further evidence of our restless planet. A fault, probably the Huasna, raised a line of volcanoes here twenty-four million years ago, the most prominent being Morro Rock, the town's preeminent landmark, rising nearly six hundred feet above Morro Bay. What you see is the volcano's core. The rest eroded away over the millennia. Thirteen sister volcanoes, similarly eroded, run southeast from Morro Bay to San Luis Obispo.

My sister-in-law and her brother own a small beach house in Morro Bay. From her cozy front deck, you can watch the surf break against the Morro Strand Beach while hummingbirds flit back and forth amid flowering bushes. The driveway is just large enough to park the Bambi. We've come here several times and have grown fond of Morro Bay and the wine country to the east in the San Luis Mountains, a triangle formed by Paso Robles to the north ("Robles," in local parlance, rhyming with "nobles"), San Luis Obispo in the south, and Morro Bay to the west. The mountainous terrain, proximity to the sea, and warmth of the inland valleys provides a wide range of terroir and thus many varieties of grapes.

An abandoned power plant with three tall smokestacks provides a second distinctive Morro Bay landmark. When the plant was constructed in the 1950s, locals abjured the smokestacks; now, the plant mothballed, they object to a Pacific Gas & Electric Company plan to remove them. The town has a pleasant, unpretentious 1970s vibe, with seafood restaurants, T-shirt shops, stores selling tourist tchotchkes, a Coast Guard Station, and a working fishing fleet that unloads its catch along the town waterfront. Sailboats swing at their moorings. The Pacific can be rough here—this is Central, not Southern California. The bay offers a last safe anchorage until northbound boaters reach Monterey. If, unlike Susan and me, you don't have resident relatives, Morro Bay State Park, on the bay itself, and Morro Strand State Beach offer camping.[9]

From Morro Bay, Highway 1 jogs inland to San Luis Obispo, merges with US 101, touches the coast again at Pismo Beach, and parts company with it once more at Arroyo Seco. The next stretch of coast is occupied by Vandenberg Space Force Base, where the United States launches its military and NASA research satellites (like Landsat) and tests its ICBMs. At Las Cruces, Highway 1 rejoins US 101, where Central California ends and Southern California begins.

When I was a kid, Southern California, especially Los Angeles, seduced me. Back then, in the 1950s and '60s, the Northwest shared a similar culture: Chevron and Union 76 gas stations, automobile-centric suburbs, ranch-style houses, supermarket parking lots, big aerospace companies, coastal mountains, the Pacific Ocean, earthquakes. We thought we were leading America into the future, without the prejudices and pomposity of "back east." But it was the unfamiliar that truly enchanted me: freeways, oil wells, Disneyland, Knott's Berry Farm, Marineland of the Pacific. Palm trees instead of conifers. Cacti instead of rhododendrons. Temperatures warm enough that it was comfortable to ocean swim. Usually we approached LA from the east, the finale to our family vacations in the Southwest, a night's drive by station wagon, towing our trailer across the Mojave Desert. I loved the anticipation as desert gave way to orange groves and palm trees and finally freeways. High-speed travel! Any time of the day! We'd arrive mid-morning at the home of our friends Jack and Muriel Whelan.[10] The Whelans lived at Portuguese Bend, north of San Pedro and south of Palos Verdes, on a bluff above the Pacific. You could walk down to the beach or over to Marineland of the Pacific, one cove north.

As the years passed, prejudice and pomposity eventually manifested themselves in Los Angeles. (Perhaps they had been there all along and I was too young to see it.) The freeways became clogged with traffic. Racial tensions erupted. The Cold War ended and the big aerospace corporations shrank. Marineland gave way to upscale housing. I imagine that the Whelans' modest house did, too. Except for business travel, LA became, at least for me, a place I had to get through to get somewhere else. And then something very Californian happened. The city reinvented itself. As Cold War industries faded, the entertainment industry blossomed beyond its traditional Hollywood core until LA could proclaim itself a world cultural center with topflight theatrical and musical venues, galleries, and museums equal to those of New York, London, and Paris and, because of movies, television, and music, with even wider reach.

Susan and I rarely approach Southern California from the desert. We arrive via US 101. After San Luis Obispo and Los Alamos but before Buellton, the highway veers sharply south and descends a defile known as Gaviota Pass. Oak trees and grassy hills give way to chaparral and rocky canyon walls.[11] Suddenly, you arrive at the Pacific Ocean.

The traditional demarcation between Central and Southern California lies only a few miles west of here at Point Conception, where the coast makes an abrupt turn east. From Point Conception to Santa Monica, this easterly lie continues. It can be disorienting, especially after a week of coastal travel: instead of the sun rising behind the mountains and setting at sea, as you face the ocean, it rises on your left and sets on your right. The east–west run also shields the shore from the prevailing, generally cooler northwest winds. The land dries out, warms up. Cacti and sagebrush displace savanna. The ecosystem here is called "sage and chaparral" and it continues all the way into coastal Mexico. It's also a major fire hazard.[12] The dense, low undergrowth provides highly combustible fuel, paradoxically a greater threat after a wet winter when vegetation has grown more profusely. The coastal Santa Ynez Mountains and Santa Monica Hills rise directly behind the beach. These are two of the seven ranges and valleys that form the Transverse Ranges that run east to west, counter to the north–south lie everywhere else on the open-ocean Pacific coast.[13] The reason for this is a tectonic larceny. The Pacific Plate, in its northwesterly slide, has twisted the San Andreas Fault system into a lazy "S" pattern. This section of coast has rotated clockwise, resulting in an east–west compression of the Peninsular Ranges so that south from Los Angeles to San Diego and into Baja California, these mountains run counter or "transverse" to their north–south norm.

We pull off the highway and get out of the car. Before us are the Santa Barbara Channel and the Channel Islands. Today there's no fog and little haze. We can see two of the islands, Santa Cruz and Anacapa, and, to the south, the first of the offshore oil platforms. In 1969, a major spill occurred at one of these, Union Oil Platform A, in the Dos Quadras Field. Public outrage led to significant restrictions on offshore drilling, rules still in place despite a number of political attempts to repeal them. The Amtrak railroad tracks run between the highway and beach. The beach is narrow and backed up by steep cliffs, both due to the rapidly uplifting land. There's a tarry tang in the air. Oil seeps naturally through the beach sand. It was here that Susan asked her question *What would we do without California?*

We continue west through Gaviota, Tajiguas, Capitan, Naples, Goleta, passing Gaviota State Park, Refugio State Beach, and El Capitán State Beach, until we finally reach Santa Barbara, a beautiful town in a hazardous place. Wildfires here pose a constant threat: In 1964, 106 homes burned; in 1977, two hundred; in 1990, five hundred; in 2008, 210; in 2009, 160. Tectonic events have twice leveled the town: an earthquake and tsunami in 1812 and another earthquake in 1925. The second quake led to a redesign of the city's architecture, which accounts for its uniform and lovely Spanish-colonial style. But the natural setting is lovely too. Mountain backdrops. Palm-shaded streets. Golden beaches. The blue Pacific.

The first offshore oilfields in the world were drilled near Santa Barbara at the beginning of the twentieth century. Oil is still being pumped (an industry not popular with locals) but the ugly derricks of the past are gone. The University of California, Santa Barbara sits here. The town is the Santa Barbara County seat. Aerospace and defense contractors do research and development here. Surfers make their home here, including several of the elder statesmen and women of the sport, but even the surfers seem conservative compared to other beach communities. The town typifies Southern California life, or at least what that life is supposed to be. Ninety thousand people live within its city limits, another 220,000 in the surrounding communities, nearly half a million in Santa Barbara County as a whole. You may begin to see the paradox. In a fault-riven desert, crisscrossed with mountain ranges, open to ocean tsunamis, short of water, subject to wildfires and mudslides, thirty million inhabitants teeter on the edge of potential environmental and tectonic disaster. And mostly they live well.

From Santa Barbara east, Highway 1 follows the base of the coastal mountains until it reaches a flat plain where Ventura, Oxnard, and Camarillo are located. At Oxnard, US 101 cuts north into the plexus of freeways that grid greater Los Angeles, while Highway 1 continues southeast along the coast to Malibu.

For wealthy celebrities, Malibu is a beach community de rigueur. Surprisingly, there's an RV park here on the steep hillside above Highway 1. Susan and I love the place, even though the sites are small, it's always crowded, and you have to cross Highway 1 to get to the beach (a life-endangering trek, especially on weekends). Our favorite site has a sweeping view: Los Angeles to the east, Catalina Island to the south, the Santa Monica Hills to the north. It even has what we've designated as "Our Personal Palm Tree." I like to sit in a camp chair on the slope and watch the Lamborghinis and surfer vans, the motorcyclists and bicycle racers that parade up and down the highway. With proximity to Santa Monica and downtown LA and wonderful art museums (the Getty, the Getty Villa, and the Los Angeles County Museum of Art, a.k.a. LACMA), I feel the pulse of a dynamic metropolis on the edge of the continent, maybe even the edge of the future. And for Susan and me, native West Coasters, frequent Highway 1 travelers, we also feel a sense of home.

THE WEST

> The West has had a way of warping well-carpentered habits,
> and raising the grain on exposed dreams.
>
> —WALLACE STEGNER

IT'S TWO DAYS SINCE a tree fell on our Bambi. We're camped in the driveway of the Camarillo, California house of our friends Dan and Gayle Coleman. The house perches on a steep ridge north of the town center. Here the Colemans keep two elderly horses, a lame dog, a horse trailer, a utility trailer, a trailerable sailboat, a car, and a pickup truck, and here they raised both of their now-adult children. To the south, we see the Pacific Ocean. Behind the house and on the neighboring ridges is chaparral scrub, so notoriously prone to California's wildfires. Below and toward the ocean is the web of townhouses, apartments, single-family homes, schools, churches, car dealerships, and shopping centers that constitutes Camarillo and Oxnard. It's not the first time we've camped here. Nor the last, either. Each time we visit, we recount our current Bambi travel adventures, which are not fully complete, it seems, until we've shared them with Dan and Gayle. A flaming Hula Hoop along Oregon's Willamette River. A midnight knocking on the door in the Mojave Desert. A tree falling at Refugio Beach. On our arrival today, Dan and I inspected the tree-fall dent in the Bambi's rooftop panel. There are other scars, too. Flecks of paint from a Montana highway's newly painted yellow line, not quite dry when we crossed it. A long scratch in the aluminum drawn by a stiff branch along the Kootenay River. A discreet undercarriage patch applied after a jack-down foul-up near Utah's Golden Spike National Monument. When we return to Seattle, the panel will be repaired. But even if we left it as it was, it would merely testify to another episode in our Bambi diary.[1]

Susan and I have covered much of the West in the last four years. If you drew our tracks on a map, it would look like a spider's web, back and forth, up and down, Seattle to San Diego, Tucson to Missoula. We've made a substantial dent in this part of the country, but not all of it. The West remains to entertain us, excite us, remind us of how fortunate we are to travel its length and breadth, to live in its embrace.

Can we generalize our impressions?

Perhaps size. More than half of the continental United States lies in the West.[2] Nowadays, Eastern friends still seem surprised when we tell them that the distance from our home in Seattle to our condo in Sun Valley, Idaho, a drive we make several times each year, is equivalent to a trip from Philadelphia to Chicago. Air travel promises to shrink the size of the West, but once you arrive in Los Angeles or Salt Lake City or Boise, you'll find yourself driving several more hours before you reach your ultimate destination.

It's much more difficult to generalize the geology of the West. If the Pacific Plate's dive under the North American Plate explains much of what you see here, separate regions often bear little resemblance to one another: sagebrush deserts, ice-crowned mountains, conifer forests, short-grass prairies, broad basins draining to no ocean, fast rivers draining watersheds the size of France, a high point (Mount Whitney) less than 150 miles from a low point (Death Valley). Western geological provinces are profoundly distinct: volcanism, basin and range, the Colorado Plateau, the Rocky Mountains orogeny. Even the climate runs to extremes: arctic cold in Wyoming, one of the hottest places on earth in Death Valley, California.

Perhaps, if the West has one thing in common, it's aridity. In his landmark 1878 *Report on the Lands of the Arid Regions*, John Wesley Powell, the first head of what became the US Geological Survey, defined the American West as beginning at the 98th meridian and ending at the Pacific Ocean.[3] Powell's biographer, Wallace Stegner, adds that while this is generally true, what really defines the West is the isohyetal line showing where twenty inches of rain falls annually. West of it, farmers must irrigate. That line begins about a third of the way across the Dakotas, Nebraska, and Kansas, then trends southwest across Oklahoma and Texas.[4] Everything about the West—its landforms, its flora and fauna, its colors, its sparseness, even the culture of its inhabitants, descends from aridity. European traditions of land stewardship worked on the eastern seaboard and in the Midwest, but failed in the arid West. Ghost towns, boom-and-bust cycles, Dust Bowl farms, denuded ranchland, drying and polluted rivers testify to those misconceptions. Even today, what Westerners demand from the land often exceeds its capacity. The great desert metropolises—Los Angeles, San Diego, Phoenix, Las Vegas, and Salt Lake City—almost certainly face a reckoning as they draw down the rivers, reservoirs, and aquifers to quench the thirst of their citizens. To quote historian Walter Webb, the West is "a semi-desert with the heart of a desert." Most people agree that the West is beautiful, but it's a beauty in peril. Land heals slowly in dry climates.

The mythologies of the West loom large in how Americans see themselves, and certainly in how the rest of the world sees America. But these mythologies, among them the lone hero at odds with society, don't serve the West well in the twenty-first century, and perhaps never did. The successful cultures here were cooperative cultures: Pueblo Native Americans, early Spanish-Americans, Utah Mormons. For every outlaw Western hero, a community of townspeople waited at the other end of a rope.

Another myth of the West—that the Federal government stole its land from the West's original European settlers—is pure bunk. Western territorial governments never owned the land, except for Texas, which entered the Union as a sovereign political state. All the others agreed to surrender federal lands when they became states. (The more troubling issue, that the original European settlers, as elsewhere in the United States, did steal the land from the original Native American inhabitants, is conveniently ignored.) For the most part, unless there is something particularly valuable in a place—oil, gold, uranium, natural gas, timber—Western states have been historically happy to let the Federal government pay for the lands' maintenance: the dams, reservoirs, irrigation projects, and highways that states could never afford on their own. It's costly to settle a desert, especially one as big as the West. Still, the battle between Westerners trying to "get back" their land continues to dominate Western politics and Western headlines and Western prejudices.

"You have to get over the color green," Stegner writes, "you have to quit associating beauty with gardens and lawns; you have to get used to an inhuman scale; you have to understand geological time."[5]

When I was a boy, each winter, my father would send off to the oil companies for the maps to the places we might visit on our next summer vacation. Often this was the first hint of the destination he had in mind. I remember the maps' arrival in the mail, crisp, unwrinkled, routes marked in felt-tip green: Carlsbad Caverns, Guadalajara, Devils Tower, Death Valley, the names themselves resonant with romance. I have a similar ritual. As winter thaws into spring, I find myself studying my computer-based maps. What places haven't we yet visited? What places would we like to visit again? How about Wyoming's Absaroka Mountains, New Mexico's Taos Pueblo, Canada's Jasper National Park? Susan and I sketch out itineraries, de-winterize the Bambi, fill its onboard water tank, confirm that its propane tanks are full, check its tires for wear and air pressure. Then, one spring day, usually early in April, we're underway on our annual shakedown cruise, perhaps around the Olympic Peninsula or over to Tofino on Vancouver Island's Pacific coast or a quick dash to the Washington wine country in Walla Walla or to Oregon's Willamette Valley.

Each year we've learned to expect change: a few more fellow travelers on the road, the campgrounds and RV parks a little more crowded. What stays unchanged is the land. It inspires us as much now as it inspired us on our first Bambi voyage, as much as when we first saw it as children. It reminds us that maybe one day it will be us, like the couple we saw in Kalaloch on our inaugural Bambi trip, walking in the rain, arm in arm, in the sunset of our lives.

ACKNOWLEDGMENTS

I wrote *Airstream Country* because of two loves: one for the American West, the other for what I call "high-end geology." My love for the American West came from my parents, John and Natalie Mathison. The summer trips they led, my brothers, sister, and me in the back seats of the family station wagon, camping trailer in tow, introduced me to the joys of the road and to the breadth and beauty of the American West. My passion for geology I owe to John McPhee from his books, published together in *Annals of the Former World*, about the revolution in geology known as plate tectonics, a theory that explains why continents and oceans are where they are, why a mountain range is in one place and not another. "High-end geology." Thank you, parents. Thank you, Mr. McPhee.

My ability as an adult to understand what I was seeing outside my car's windows I owe to the *Roadside Geology* series of books, published by Mountain Press of Missoula, Montana, each for a specific Western state, the chapters organized by major highways. Thank you, *Roadside Geology* authors Felicie Williams, Halka Chronic, Marli Miller, Magdalena Sandoval Donahue, Lucy Chronic, Darrel Cowan, Paul Link, Shawn Willsey, David Alt, Don Hyndman, Robert Thomas. Thank you also to the unnamed geologists selected by the University of New Mexico Press to review *Airstream County*'s science and who generously shared new geological theories that made the book more up to date.

Most writers have a foundation upon which they build their work. For me this has been Priscilla Long, poet, essayist, historian, and also my teacher, mentor, and writerly confidante, who encouraged me to submit my manuscript to the University of New Mexico Press. Thank you, Priscilla. Thank you also to my classmates in Priscilla's Advanced Short Forms class, who, over two decades, became companions and cheerleaders for each other's writing.

Thank you, University of New Mexico Press, especially Brenton Woodward, who persuaded the other UNMP editors to accept the manuscript and who has been an unfailing source of good advice and encouragement. Thanks, too, to UNMP production manager James Ayers and production editor Anna Pohlod, who have worked to pull the book together. Thank you, Katherine Harper, my manuscript editor, who, with a gentle but firm hand and with extraordinary attention to detail, made the book better.

I want to thank my brothers, Charles and Duncan Mathison, and their spouses, Patsy Mathison and Karen Georgatos, who, like Susan and me, have returned to trailering in their autumn years. They have joined us on journeys, created new adventures, and, with affection for each other and much laughter, recalled past journeys back to life.

Thank you to my sister, Charlotte Guyman, and my brother-in-law, Doug Guyman, who, over twenty-five years, have encouraged and celebrated my writing and tolerated having a memoirist in the family.

Thank you, John Mathison, our son, for joining us on Bambi trips, mountain tent–camped outside our door. Your presence always makes our family feel complete.

Lastly, thank you, Susan Cole. When Susan packs a shelf in our sixteen-foot Airstream Bambi, it has the precision and beauty of a Vermeer painting. Whether a trip is six days or six weeks, she knows exactly what she brought along, where she stored it, and exactly when we'll need to shop for provisions. She remains unfailingly curious about what lies beyond every turn in the road, and is always ready to help back the Bambi into a tight campsite. When we return home, all I need is to tell her a day and a year, and Susan, checking her detailed travel journals and spreadsheets, can tell me where we were, what we did, and (often) what we had for lunch. I can imagine no better companion for an Airstream journey and for a journey through life.

NOTES

Chapter 2

1. Perhaps the entire continent isn't up-thrusting, but subduction of the Pacific Plate under this edge of the continent is up-thrusting the Olympic Mountains.
2. Kettles, drumlins, and esker rocks are the leitmotivs of glacial activity. Kettles are depressions in the glacial outwash; drumlins are elongated, spoon-shaped hills shaped by the ice; esker rocks are rocks carried by and subsequently deposited by the ice.
3. If you're of a certain age, the name "Bambi" evokes the Walt Disney animated movie of the same name. The film, released in 1942 and then rereleased in 1947, 1957, 1966, 1975, 1982, and 1988, followed by its issue as a home video in 1989, was a coming-of-age milestone for most of us mid-to-late-twentieth-century kids—especially because of the death of Bambi's mother at the hands of human hunters, death in those days rarely depicted in children's movies. The Airstream Bambi, however, was not named after the iconic Disney fawn. In 1961, as the company prepared to introduce a lightweight single-axle trailer, founder Wally Bynum had just returned from his Airstream caravan tour to sites across Africa. While in Angola, he learned of a miniature deer known for its strength and sure-footedness. In the Umabundu dialect of the Bantu language, this animal was called *O'Mbambi*. Bynum concluded that a shortened version of the name was a natural for the new trailer. As Director of Sales C. H. Manchester put it in a letter to Airstream dealers, "Bambi means speed. Bambi means stamina."

Chapter 3

1. The San Juan volcanoes are not only older than the Cascade volcanoes, but also represent a different type of volcanism, in which enormous subsurface magma chambers empty, leave a hole in

the ground, and erupt on a scale modern humans have never recorded.

2. Geology, like all science, is dynamic. New data have yielded new theories since Williams and Chronic published *The Roadside Geology of Colorado*. A now-vanished plate, the Farallon, once existed between the Pacific and North American Plates. This theory suggests that the rise of the Rocky Mountains and the Silverton volcanic eruptions were both triggered by the Farallon's behavior as it completely subducted under North America. At first, the subduction angle was moderately steep, creating the plutons that underlie California's Sierra Nevada, but between seventy million and forty-three million years ago, the angle of subduction flattened, either due to an increase in the rate of subduction or subduction of a buoyant ocean plateau. Compression of the earth's crust associated with this low angle of subduction led to the formation of the Rocky Mountains during what is known as the Laramide Mountain Building event. Subsequently, subduction slowed. The Farallon's angle of subduction increased, causing the plate to sink into the hot mantle and triggering the explosive and extensive San Juan/Silverton volcanism.

3. Felicie Williams and Halka Chronic, *Roadside Geology of Colorado*, 3rd ed. (Missoula, MT: Mountain Press, 2014), 254.

Chapter 4

1. At this writing (2023), the state parks' problem is drought. California's governor, Gavin Newsom, has ordered all state agencies, including the state parks, to conserve water. Showers in camp facilities have been secured. Sewage dump stations at campgrounds have been shut down.

2. For most of humankind's existence, people have lived in close quarters with their biological waste; only in the last century were significant numbers of us liberated from having to dispose of it. RV travel, however, has brought waste disposal back into the personal domain. Dump stations are more or less the same everywhere: They have a catch basin with a hole in the ground that leads to the local sewer or septic tank system. The hole is capped with a metal lid and a foot pedal so you don't have to touch the cap with your hands. Black-water tank connections are underneath most RVs, usually in the rear and, by industry convention, on the vehicle's left-hand side. You need to maneuver close to access

the dump station. In our case, I can't see the relative positions of trailer and catch basin from my driver's seat, so Susan stands outside the car and directs me where to steer. The Bambi must be close enough that our hose—expandable, flexible, six feet long, six inches in diameter, much larger than a garden hose—will reach the dump station connection. Then the fun begins. I open the Bambi's rear storage door and haul out the dump-station kit. I don latex disposable gloves. I fit an L-shaped plastic connector to the dump hose that is stored separately in a tube forward of the Bambi axle. I open the cap on the dump connection (hoping that the valve hasn't leaked), attach the hose, and lead it to the dump station hole (hoping that the previous user has rinsed off the catch basin, not always the case). Susan stands on the foot pedal to open the cap while I fit the L-shaped connector into the hole. Then I open the T-shaped handle that operates the Bambi dump valve. Because the hose connector on the Bambi end is clear plastic, I can monitor its flow (trying not think about what I'm seeing). When the flow stops, I reverse the procedure: disconnect the hose, wash everything, re-store the gear, throw away the latex gloves, douse my hands with an antibacterial solution. Meanwhile Susan adds a deodorizer/sanitizer to the black-water tank. For most RVers, dumping seems to be solely a male vocation, with most female partners waiting de-murely and distantly. Susan helps. Barring problems, the evolution takes fifteen minutes. Often, of course, you have to wait your turn behind other RVs. On a Sunday afternoon, with campers returning home, the lines can be long. Anticipation doesn't make this task any more pleasant.

3. Every campground has rules for generator operation. In National Parks, they are restricted to a couple of hours in the morning and late afternoon; state parks and National Forest campgrounds allow more hours. In no case are generators permitted after ten p.m. or before eight a.m. Very polite RVers don't run their generators at all in a campground, saving them for wilderness sites. (Susan and I number ourselves among these.) Apologetic RVers run them only briefly. Renegades ignore the rules.

Chapter 5

1. That these fine old names, Ahwahnee and Camp Curry, had to be changed, seemed a shame. The old park concessionaire, who lost the concession contract, had trademarked them. Since our visit,

I'm happy to report, the names were returned to the current concessionaire.

2. "You can't wait for inspiration. You have to go after it with a club" (Jack London).

3. According to the *Encyclopaedia Britannica*, the Central Valley produces about one-quarter of the nation's crops on less than 1 percent of its land.

4. The Sierras technically lie in between the Central Valley and the Basin and Range province and can be thought of as part of the Ancestral Cascade Arc. During the Cenozoic Era, numerous small eruptions occurred along the Sierran fault zone about the same time as the Basin and Range started forming.

5. The plutons that underlie the Sierras owe their origin to the subduction of the Farallon Plate, as described earlier in the discussion of the rise of the Rocky Mountains.

Chapter 6

1. This fascinating tidbit is discussed on the website of the Joshua Tree Genome Project (https://joshuatreegenome.org/archives/2016/03/what-is-the-deal-with-joshua-trees-and-yucca-moths/).

2. A *mole* is a Mexican sauce and it's served in a wide variety of tastes, but usually contains a fruit, chili pepper, and nut, sometimes even chocolate. Not your neighborhood Taco Time.

3. The white limestone we see in Oak Creek Canyon is the same limestone found at the top layer of the Grand Canyon.

Chapter 7

1. Here a trailering digression may be appropriate. Hooking up a modern travel trailer is not as simple as hooking up a U-Haul trailer. Like the U-Haul, the Bambi has a coupler that drops over the tow vehicle's hitch ball, a pair of safety chains (in case the hitch breaks—it does happen), and an electrical connector for the trailer lights. But in addition, most travel trailers have load-balancing antisway bars, electric brake safety lanyards, and a larger electrical connector that provides power to the trailer brakes, batteries, and side and taillights. Sway bars come in pairs and extend from underneath the car hitch in a Y to either side of the trailer tongue. Because the bars connect to the hitch, the hitch is heavy and bulky—a shin-banger when installed in the car receiver, so, when

not towing, Susan and I remove the hitch. When we want to hook up, we reinstall and pin the hitch in the car's hitch receiver, back the car up so the ball is directly below the hitch coupler, lower the trailer tongue receiver over the ball, and click the coupler locking lever into place. We jack up the front end of the trailer with the trailer tongue–leveling jack so that the sway/spring bars will slide over flanges on either side of the tongue. We slip the bars over the flange and clamp them into place. Then we lower the tongue jack so that the car bears the trailer weight. This tensions the bars and distributes the trailer weight between the car's front and rear axles so that the tow vehicle is level (load-balancing). On our hitch, these bars also restrain "fishtailing," the tendency of the trailer to weave back and forth in an exaggerated fashion as the car steers right and left (sway control). The brake safety lanyard hooks to the car hitch so as to trigger the electric trailer brakes if the trailer becomes un-hitched while underway. There's nothing particularly complicated about this—an experienced RVer can connect it all in five minutes. But Susan and I have learned to cross-check everything. It's easy to forget the electrical connector or the brake lanyard or the safety chains in the rush to get underway.

2. While environmentalists—at the time a nascent movement—failed to halt construction of the Glen Canyon Dam, the uproar they raised did stop other dams, some of which would have inundated portions of the Grand Canyon.

3. In Zion Canyon itself, the Gray Cliffs are Cretaceous Age and the Pink Cliffs are actually Tertiary (the beginning of the Cenozoic Era—the age of mammals).

Chapter 8

1. As I write this, in 2023, it appears that is exactly what has happened.

2. The viscosity of a volcano's lava determines its character: more silica and more gas in their magma yields thicker, slower-flowing lava (usually andesite and rhyolite) and thus a more explosive eruption. Newberry demonstrates a common progression for stratovolca-noes, beginning with less-explosive basalt flows, then becoming more explosive as the lavas become more silicon-rich.

Chapter 9

1. Scientists only began to understand supereruptions and

supervolcanoes in the 1970s through the 1990s, by study of the Valles Caldera in New Mexico and others elsewhere in the world.

2. As a kid, in 1956, I attempted to photograph the boiling mud with a Brownie box camera, an unsuccessful venture. All I photographed was steam. The boiling mud is technically "mudpots" and not a volcano at all.

Chapter 10

1. Small glaciers still exist, but these formed during the "Little Ice Age" that ran from the 1400s to 1850. Like many glaciers in our warming world, these may soon be gone.

2. NOLS has done much to improve outdoor skills and safety, wilderness medicine, and environmental ethics. The conservation program Leave No Trace is co-managed by NOLS.

3. Several years later, on another Bambi trip, we will pass closer to the Wind Rivers and we will be impressed.

4. The area's occasional French place names originate from the French fur-trade trappers of the early nineteenth century. There's a lot of history here: geological, biological, Native Americans, mountain men, ranchers. Susan and I have just begun to comprehend how little about Wyoming we really know.

Chapter 11

1. This type of collapse is so common in Iceland that Icelanders have invented a word for it: *Jökulhlaup*. In Nepal, because of global climate change, several mountain communities are now threatened by ice-dam Jökulhlaups.

2. Rick Bass, "Fires," *In the Loyal Mountains* (Boston: Houghton Mifflin, 1995), 37.

3. Several years from now, in an expensive Bay Area RV park, we'll discover that our "antitheft" straps are no barrier to thieves when our two road bikes are stolen, straps clipped, the car less than ten feet from where we sleep in the Bambi.

4. The Glacier National Park website seems attuned to this. A video about the Apgar campground features families surrounding a campfire, roasting marshmallows, layering melted marshmallow and chocolate into graham-cracker sandwiches.

Chapter 12

1. During its eight-hundred-mile run, Highway 1 goes by different names: the Pacific Coast Highway (PCH), the Shoreline Highway, and local names along the way such as Lincoln Boulevard and Sepulveda Boulevard.
2. As always with California's geology, the story is more complicated. Well before the most recent Ice Age, over millions of years, the valley basin filled with ten thousand to fifteen thousand feet of marine and continental sediments and, more recently, eighty thousand to 125 thousand years past, underwent minor marine transgressions as well as erosion of adjacent uplifted mountains, depositing additional layers as gravel, sand, silt, and clay.
3. Susan and I cart around two inflatable kayaks and an inflatable paddleboard with their respective life jackets and paddles. Space is dear in the MDX and the Bambi. At times this stuff has seemed too much for too little return, but on the several occasions when we've launched them—San Diego's Mission Bay, Idaho's Redfish Lake, Washington State's Banks Lake, here at Cannery Row—our voyages have been particularly memorable. So, for the time being, as long as we're fit enough, the kayaks and paddleboard will stay onboard.
4. The song, featuring the lyrics "I'd like to teach the world to sing / In perfect harmony / I'd like to buy the world a Coke / And keep it company / That's the real thing" proved so popular that it became a radio hit as well.
5. In the interest of full disclosure, I once recommended a Big Sur drive to my longtime dentist and friend Al Kariya, who subsequently drove its entire length in dense fog.
6. The dominant movement in the area is strike/slip rather than vertical.
7. Patty Hearst, William's granddaughter and onetime kidnap victim of the Symbionese Liberation Army, reports that she used to hide behind the swimming pool's Greco-Roman statues while spying on Castle tourist groups.
8. A website called earthquaketrack.com lists each day's earthquakes worldwide. Look for Central California. The number each day (albeit most are small) is thought-provoking.
9. There are several private RV parks and motels in Morro Bay and also farther north in the attractive coastal community of Cambria.

10. The Whelans had shared a house with my parents when, post–World War II, they lived in São Paulo, Brazil.

11. Gaviota Pass appears in two movies: *The Graduate*, where Dustin Hoffman's character, supposedly southbound, goes through the northbound-only tunnel. In *Sideways*, the Paul Giamatti and Thomas Hayden Church characters go through the tunnel in the correct, northerly direction.

12. As I write this, wildfires are raging along this stretch of coast.

13. The north shore of the Olympic Peninsula also runs east–west, but technically that coast borders the Juan de Fuca Strait rather than the Pacific Ocean.

Epilogue

1. In the fall of 2017, as wildfires raged just north of the Colemans' Camarillo home, Dan sent us an email: "Your Bambi campsite is still here."

2. In the 1960s, my parents' Pennsylvania friends Bob and Esther Baldwin offered to meet us "halfway across," an offer they withdrew when my dad pointed out that "halfway across" was Billings, Montana.

3. Powell was one of the first to advocate for strict conservation of water resources in the arid West. "There is not enough water to irrigate all the lands," he remarked at a congress of farmers and developers in 1893. "I tell you, gentlemen, you are piling up a heritage of conflict and litigation over water rights, for there is not enough water to supply the land."

4. Stegner wasn't writing about the far corner in the northwest of the West, that thin, rained-upon band on the Washington and Oregon coast west of the Cascade Mountains.

5. Wallace Stegner, "Living Dry," *Where the Bluebird Sings to the Lemonade Springs: Living and Writing in the West* (New York: Penguin, 1992), 54.